SUSTAINABILITY & HISTORIC PRESERVATION

Toward a Holistic View

SUSTAINABILITY & HISTORIC PRESERVATION

Toward a Holistic View

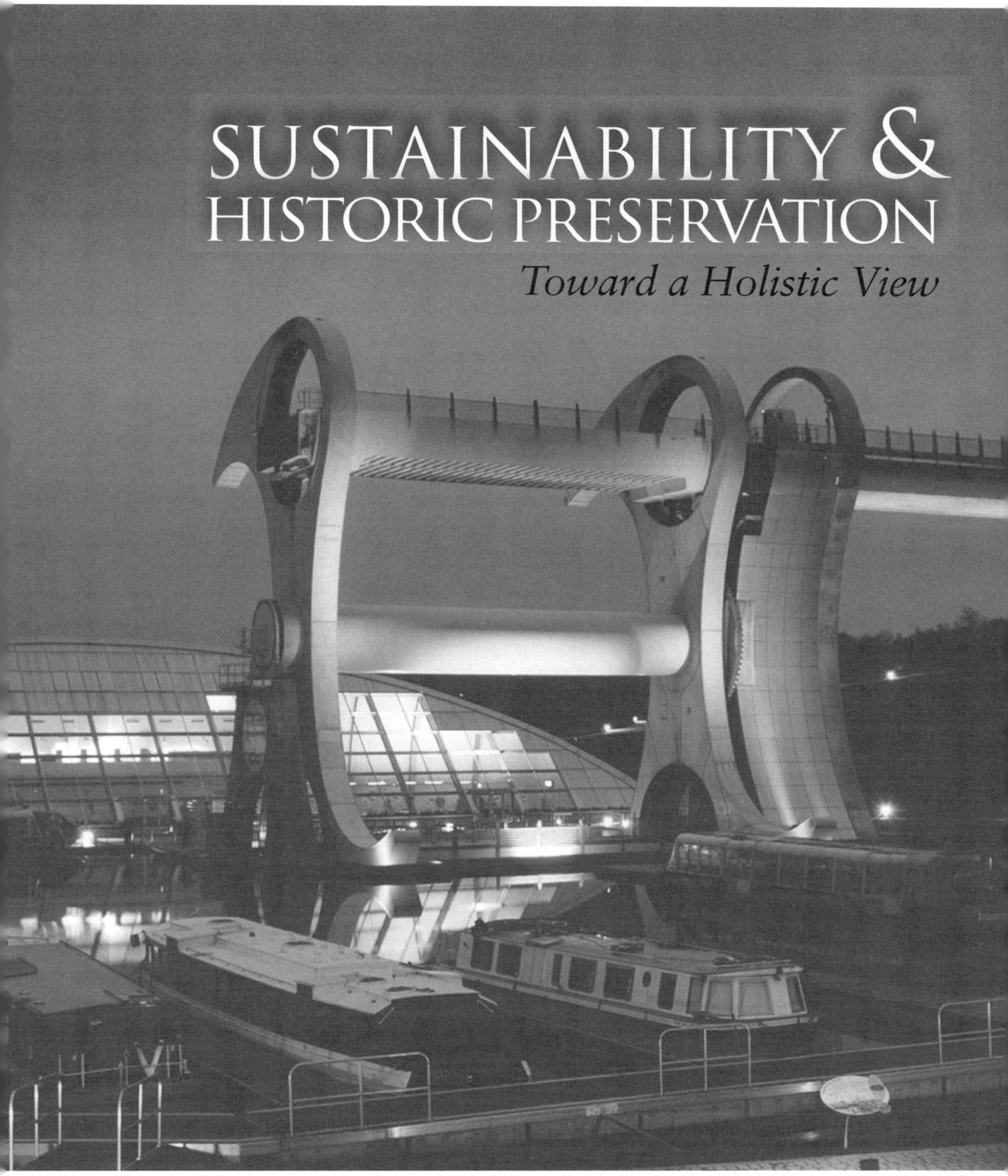

edited by RICHARD LONGSTRETH

UNIVERSITY OF DELAWARE PRESS
Newark

Published by UNIVERSITY OF DELAWARE PRESS
Co-published with The Rowman & Littlefield Publishing Group, Inc.
4501 Forbes Boulevard, Suite 200, Lanham, Maryland 20706
www.rlpgbooks.com

Estover Road, Plymouth PL6 7PY, United Kingdom

British Library Cataloguing in Publication Information Available

Library of Congress Cataloging-in-Publication Data Available
ISBN 978-1-61149-337-5 (cloth : alk. paper)

The paper used in this publication meets the minimum requirements of American National Standard for Information Sciences—Permanence of Paper for Printed Library Materials, ANSI/NISO Z39.48-1992.

Printed in the United States of America

Contents

PREFACE

Richard Longstreth

This book is the outgrowth of the Sixth National Forum on Historic Preservation Practice, "A Critical Look at Sustainability and Historic Preservation," held at Goucher College in Towson, Maryland, on March 19-21, 2009. The basic objective of these conferences is to provide an all-too-rare opportunity for people working the in the preservation field to hear and discuss current issues that are complex and often the source of disagreement or misunderstanding.[1] The exchanges are conducted on "neutral ground," as it were, in the halls of the academy rather than under the aegis of a preservation organization so that all participants may feel unencumbered in expressing their views.

A number of academic institutions offering graduate instruction in the historic preservation field serve as co-sponsors, including Boston University, Columbia University, George Washington University, Goucher College, the University of Cincinnati, the University of Delaware, the University of Florida, the University of Kentucky, the University of Maryland, the University of Minnesota, the University of Oregon, and the University of Southern California. Equally important for the sixth forum was co-sponsorship by the American Institute of Architects Historic Resources Committee and the National Institute of Building Sciences. National Park Service personnel from the Hampton National Historic Site made special contributions to the conference. Goucher College provided an ideal setting and admirable support through the offices of Megan Cornett, Noreen Mack, and Craig Henderson.

The papers presented were chosen from a large number of abstracts sent in response to a widely distributed call. The review committee consisted of David Ames, Kenneth Breisch, Claire Dempsey, Robert Mack, Hugh Miller, de Teel Patterson Tiller, Jeffrey Tilman, Richard Wagner, and myself. The selection process was based solely on the merits of the proposals received rather than being shaped in part by a concern for comprehensiveness or for attaining a balanced program in terms of subject matter, viewpoint, speaker background, or geographic areas covered. The committee also reviewed an initial draft of the papers. The proceedings were printed as David Ames, ed., "A Critical Look at Sustainability and Historic Preservation," 2009. Out of the twelve papers presented, seven were developed further for this anthology. My concluding remarks at the conference have likewise been recast as the final chapter. Richard Wagner has prepared an introductory chapter expressly for this volume.

I am grateful to David Ames and Richard Wagner for their overall coordination of this project, which has made my job easier in innumerable ways. Special thanks go to the authors for producing insightful work and for responding to my editorial queries and comments. Each chapter has been prepared so that it may stand on its own, but I have sought to minimize redundancies and otherwise structure the material so that it contributes to a larger, coherent whole.

This book makes no attempt at being comprehensive in scope. Quite to the contrary, its contents are but a small sampling of what could be many more contributions to a vast and diverse subject. But the chapters do suggest the wide spectrum of topics that relate historic preservation and sustainability. The authors' range of professional expertise, concerns, outlooks, and opinions all indicate how rich the exploration of this relationship has become. Collectively, their work makes clear that while concerns of sustainability are changing and will continue to change the nature of preservation, the opposite is also true. The perspective no less than the practices of historic preservation can contribute much to enhance the pursuit of sustainability. The dynamic is one of mutual enhancement. An equally important conclusion that can be drawn from these chapters is the key role played by people – not just those working in a professional capacity, but those in the communities involved. Neither sustainable nor preserved environments can materialize on a substantial scale without broad-based public participation.

The book is organized to present its varied subject matter in a logical sequence. Richard Wagner provides a concise overview of the rise of concern over sustainability within the larger context of the environmental movement, concluding that some preservation practices may have to change if these two spheres are to be effectively integrated. Thomas King explores an equally broad field, drawing from a spectrum of his international projects to underscore not only some of the fundamental differences that exist between Western and non-Western cultural outlooks, but also a conceptual framework that can effectively bridge that divide.

Shary Page Berg addresses the crucial need to preserve landscapes in a holistic way that encompasses both environmental and cultural spheres. Her chapter further emphasizes the vital part community values play in identifying significant places on a local level, just as King does on a global level. While Rebecca Crew's focus is on the natural environment, she demonstrates how closely components of that environment are tied to our cultural and regional perspectives and how telling the interface can be in urban as well as in rural environments. Valencia Libby's case study reveals the enduring value of a private park that has become an urban oasis and how its impact has been extended to the lives of the underprivileged people who now live in the vicinity.

Karana Hattersley-Drayton continues with the theme of relationships between nature and culture by analyzing the significance of the extensive irrigation system that has served both Fresno and surrounding agricultural areas of California's San Joaquin Valley for well over a century. Meisha Hunter also focuses on water as infrastructure, but in an entirely urban framework, suggesting that such systems may serve useful purposes for far longer periods than generally occurs with common practices.

Liz Robinson focuses on an innovative program of retrofitting modest Philadelphia row houses that is both more energy efficient and hospitable for their ongoing use by low-income residents, demonstrating in the process how conserving these habitations can be accomplished on a cost-effective basis. My remarks are intended to reiterate some of the issues raised by the other authors, but also to raise additional ones, underscoring the great array of landscapes than can and should benefit from a holistic integration of sustainability and historic preservation objectives. The ominous threat of global warming can yield surprisingly beneficial results if we address the myriad challenges it presents with respect for the past no less than with concern for the future.

Note

1 Issues examined in the five previous meetings were historical significance, integrity, design, cultural landscapes, and authenticity. A refined version of the first conference was published as Michael Tomlan, ed., *Preservation of What, for Whom? A Critical Look at Historical Significance* (Ithaca, N.Y.: National Council for Preservation Education, 1999). Publications developed from the third and fourth conferences are: David Ames and Richard Wagner, eds., *Design & Historic Preservation: The Challenge of Compatibility* (Newark: University of Delaware Press, 2009); and Richard Longstreth, ed., *Cultural Landscapes: Balancing Nature and Heritage in Preservation Practice* (Minneapolis: University of Minnesota Press, 2008). Several of the papers presented at the fifth conference were published in *CRM: The Journal of Heritage Stewardship*: David Lowenthal, "Authenticities Past and Present" and Robert Garland Thomson, "Authenticity and the Post-Conflict Reconstruction of Historic Sites," *CRM* 5 (winter 2009): 6–17 and 64–80, respectively; David G. DeLong, "Changeable Degrees of Authenticity" and Frederic L. Quivik, "Authenticity and the Preservation of Technological Systems," *CRM* 5 (summer 2008): 28–38, respectively; and David N. Fixler, "Material, Idea, and Authenticity in Treatment of the Architecture of the Modern Movement," *CRM* 6 (winter 2009): 8–22.

Finding a Seat at the Table: Preservation and Sustainability

Richard Wagner

It is extremely difficult to read anything today on the environment, culture, or economics without encountering some discussion of sustainability. The subject pervades advertisements, entertainment, and popular culture, as well as boardrooms of corporations, centers of higher education, and think tanks. To be sustainable is to be in step with society's desire to co-exist with nature. Not to be sustainable is considered old-fashioned and out-of-step. To many, sustainability is the issue of the twenty-first century. It suggests that we have finally realized that we can no longer continue to consume resources at an ever-expanding pace without causing irreparable damage to our planet.

Coinciding with the worldwide economic slowdown of the last two years, the push for sustainable living has intensified manyfold, including the practice of historic preservation. National preservation organizations and federal agencies place sustainability at the top of their agendas at meetings and conferences. Preservation architects have become concerned with making historic buildings sustainable, attempting to marry the Secretary of the Interior's Standards for Rehabilitation with the various green building-rating systems to garner tax credits for their clients. State and local preservation offices are concerned with meeting new standards of energy consumption and the rehabilitation of historic properties developed for their own jurisdictions.

Historic preservation, particularly in the United States, is a relative newcomer to sustainability. On the other hand, the conservation movement in this country and abroad has a long tradition of concern for the effect of human activity on the environment. Demographers, environmental scientists, and even economists have examined the effects of growth on natural resources for well over three hundred years. In the eighteenth century, the Reverend Thomas Malthus was concerned that the population growth fueled by the beginnings of the Industrial Revolution would outstrip humanity's ability to feed itself. In fact as long ago as ancient Rome, writers speculated on the ability of the planet to support the growing population: "We have become a burden to our planet. Resources are scarce and soon nature will no longer be able to satisfy our needs."[1]

Origins of the Modern Sustainability Movement

Many trace the birth of the sustainability movement to 2004 when Al Gore presented his slide show on global warming at the premier of Roland Emmerich's film, *The Day After Tomorrow*, which depicted a new ice age caused by the effects of global warming. Gore, a former senator and vice president, had long been interested in environmental issues. Emmerich's film, which grossed over one-half billion dollars worldwide, gave popular voice to that concern. Within two years, Gore had produced his own, more serious film, *An Inconvenient Truth*, which engendered enormous interest throughout the world in climate change and its effects on our environment.

While these two films can be seen as raising the awareness of the general population on global environmental issues, scholars and diplomats often trace the current interest in sustainability to the Kyoto Protocols. Named for the 1997 host city of the United Nations Framework Convention on Climate Change, the objectives of the Protocols was to develop binding agreements to reduce greenhouse gases by industrialized countries, while allowing non-industrial countries to increase emissions, and thus be able to grow economically. Thus the Protocols explicitly recognized the link between economic growth and increased consumption of energy resources. To balance global growth between the industrialized and emerging countries, the Protocols popularized the concept of carbon cap-and-trade, which attempts to establish an economic value for a "carbon credit."[2] The goal is to use market forces to eventually drive commercial enterprises toward sustainable practices.

In effect, a carbon credit market attempts to create a framework to equate emissions to a monetary value that is subject to supply-and-demand, potentially creating a market far larger than the world's stock exchanges or futures markets. As William Nordhaus has pointed out, the carbon credit market will become effective when the price of a carbon credit is high enough to alter economic behavior.[3] He suggested that if the price of carbon credits was high enough to attract significant trading, it would achieve a number of goals. First, it would provide consumers with knowledge about what goods and services are high in carbon content, which would cost more than those with low carbon footprints. Second, a high carbon cost for goods and services would provide producers with incentives to move from sources of energy high in carbon, such as coal or oil, to less expensive ones, such as wind or nuclear power. Third, the higher the price of a carbon credit, the more economic stimulus this would provide to create lower carbon products and processes. Finally, Nordhaus suggests that a carbon price of $30 per metric ton would be necessary to jump-start the carbon credit markets, which in turn would fuel new technological and economic expansion.

An alternative to trading in carbon credits, with its reliance on market forces and its susceptibity to fluctuations in price (favored by the Kyoto Protocols) is a tax on the consumption of carbon.[4] A carbon tax is one on a transaction that is considered by society to have a negative externality; that is, a transaction that produces a negative indirect effect.[5] Thus, a tax on releasing carbon dioxide into the atmosphere is similar to "sin tax" on tobacco and alcohol imposed in part to deter consumption of these products by citizens. Arthur Pigou, an early twentieth-century British economist, suggested that taxing goods associated with a negative externality is the best method of determining their true cost to society.[6] He even lent his name to a tax that attempts to include negative externalities in the price of goods—a Pegovian tax. Many economists, including Paul Volker, who recently emerged from retirement to advise the Obama administration on the economy, agree that imposing a Pegovian tax on large carbon producers such as oil and coal companies might be better than waiting for the free market to drive up prices as Nordhaus advocates. To date, a number of Scandinavian countries have implemented carbon taxation with mixed results on the effect on carbon footprints. In 1993, Bill Clinton proposed a national BTU tax be implemented in the United States.[7] Backed by Vice President Al Gore, this proposal was thought to contribute to the Democrats losing seats in both the House and Senate during the following year's by-election.

The birth of the sustainability movement also has its roots in the so-called oil crisis of the 1970s. In early October 1973, Egypt and Syria launched coordinated attacks on Israel on Yom Kippur, the holiest day of the Jewish year. In mid-October, in response to aid provided Israel by the United States and Western Europe, the Organization of Petroleum Exporting Countries (OPEC) voted to cut production, triggering oil and gas shortages in many countries. That measure, in turn, gave rise to a number of ideas and programs to cut energy consumption, including the Emergency Highway Energy Conservation Act (1974), which reduced maximum speeds on federal or federally funded highways to 55 mph (1974), Corporate Average Fuel Economy (CAFE) Standards for automobiles (1975), and the Energy Policy and Conservation Act (1975), which set the first energy consumption standards for appliances and led to formation of the U.S. Department of Energy (1977).

In the same year as the Yom Kippur War, a well respected but little known British economist published *Small Is Beautiful: Economics As If People Mattered*. In it, Ernst Fredrich (E. F.) Schumacher, the recently retired chief economic advisor to the U.K. National Coal Board, argued that modern economic growth was ultimately unsustainable since it treated natural resources as income rather than capital.[8] He made economic arguments supporting a number of popular movements of the time, including the back-to-the-earth movement of the late 1960s, and transferring technologies from the first to the third world, as the industrial west and undeveloped countries were then known. He also attacked the notion that big was better, substituting the concept that cottage industries were the best way to ensure our long-term future.

The year before *Small Is Beautiful* and the Yom Kippur War, the Club of Rome published *The Limits to Growth*, which warned of the effects of unrestrained growth on the earth's resources.[9] Using a relatively simply model to examine the interaction between world population growth, increased industrialization and pollution, food production, and resource depletion, the book echoed the predictions of Malthus almost 200 years previously.[10] The authors of *The Limits to Growth* admitted that the inputs to the model, particularly the nature and extent of global natural resources, were only estimates. Nonetheless, the outcomes were presented as a mathematical model of the effects of the world's growing populations on natural resources. However, because of the inexact nature of the inputs, a number of thoughtful criticisms of the model were published in scholarly journals and the popular press. Most of the critiques complained that the model included exponential growth in world population and pollution, but did not account for technological innovation and the discovery of new resources to accommodate that growth.

Prior to the 1970s, one can cite a number of key organizations and publications from the post-World War II period that contributed to the foundation of today's sustainable movement, including the Friends of the Earth (1970), the Environmental Law Institute (1969), *Silent Spring* (1962), and the Nature Conservancy (1951). Friends of the Earth, founded at the height of the hippie era, is a grassroots organization operating in over seventy-five countries "defending the environment and championing a healthier and just world, which pulls no punches, and speak sometimes uncomfortable truths to power."[11] Over the years, the Friends have been an effective, if sometimes strident, voice for environmental policies. In contrast, the Environmental Law Institute, founded just a year earlier, is a scholarly research and educational organization devoted to improving environmental laws and policies worldwide. Known primarily for its flagship publication, *Environmental Law Reporter*, the Institute publishes a wide range of papers, books, and articles on environmental policy and sustainability.

Rachel Carson, who is often recognized as the godmother of the environmental-sustainability movement, was born in 1907 and grew up on a farm outside of Pittsburgh. In 1935, having earned an undergraduate degree in English from the Pennsylvania College for Women and a graduate degree in zoology from Johns Hopkins University, Carson went to work for the U.S. Bureau of Fisheries, where she stayed until after the publication of her second book in 1950, *The Sea Around Us*. Its popular and critical success allowed Carson to support herself as an author.[12] Twelve years later, her seminal work, *Silent Spring*, was published, quickly reaching the top of the *New York Times* best-seller list as well as becoming a Book-of-the-Month Club selection. In it, Carson documented the detrimental effects of pesticides on the environment, particularly birds. She demonstrated that concentrations of DDT, then widely used to control mosquitoes, caused bird eggshells to thin, thus breaking easily before chicks could be hatched. As one might suspect, Carson and her book were widely attacked by companies with vested interest in pesticides, but within a few years, the evidence presented by her caused many of these companies to rethink their products, developing new ones that were effective as well as neutral to the environment.

Between the beginning of the twentieth century and the end of World War II, the world's attention was largely consumed with national and economic survival, leaving little time for concern with sustainability or the environment. During the last decade of the nineteenth century, however, two key organizations were founded that may be considered the foundation of the modern sustainability movement— the National Trust for Places of

Historic Interest and Natural Beauty in Great Britain, and the Sierra Club in the United States. The British National Trust was founded in 1895 by Octavia Hill, Robert Hunter, and Canon Hardwicke Rawnsley to preserve "for the benefit of the Nation of lands and tenements [including buildings] of beauty or historic interest and, as regards lands, for the preservation of their natural aspect, features and animal and plant life".[13] Hill, a protégé of John Ruskin, was a social reformer who was concerned with improving tenement housing in English cities as well as the preservation of the countryside. Hunter was an attorney who became interested in the mid-nineteenth century in saving common lands from enclosures.[14] Hardwicke was a clergyman from the Lake District concerned with its rapid development at the end of the nineteenth century for vacation houses by the newly affluent middle class, who could rapidly travel to a residential retreat on the country's ever-expanding railroad system. Unlike its much younger American counterpart, the British National Trust has always been interested in the preservation and conservation of the countryside as well as buildings.[15] Today, it owes over 1.5 percent of the land area of England, Wales, and Northern Ireland, has 3.5 million members, and an operating budget of nearly $450 million.[16] In common with most organizations concerned with conservation and preservation, the British National Trust joined the sustainability movement more than a decade ago with a broad range of initiatives ranging from sustainable historic buildings and towns, rural sustainability, sustainable food production, and education on sustainability in grade schools, to name but a few programs.

The Sierra Club was founded in San Francisco by John Muir, then at the height of his authority as America's first great conservationist. Its mission is "To explore, enjoy, and protect the wild places of the earth; To practice and promote the responsible use of the earth's ecosystems and resources; To educate and enlist humanity to protect and restore the quality of the natural and human environment; and to use all lawful means to carry out these objectives."[17] Originally focused on bringing western wilderness under the protection of the United States government, the Sierra Club slowly grew from a few hundred members to 15,000 in 1950. As the nation's awareness of environmental issues developed in the 1960s, the Sierra Club published *The Population Bomb* by Paul Ehrlich, another in a growing number of publications warning of the planet's limited resources to support its ever-expanding population. By the 1990s, the organization had become deeply involved in environmental issues such as controlling greenhouse emissions, developing green energy sources, and green transportation, as well as moving into new areas such as in safeguarding centers of human habitation from the coming effects of climate change.[18]

Defining Sustainability: The Brundtland Commission and Smart Growth

Sustainability means different things to different people. Some see it as a back-to-the-earth movement with cottage industries replacing multinational corporations. Others see it as an effort to produce green buildings and cities, cutting energy consumption to achieve the holy grail of a zero carbon footprint. Still others see sustainability as preserving rapidly vanishing cultural diversity of the world, stemming the tide of globalization on societies. While sustainability may entail these spheres and others as well, many authorities define sustainability using the United Nations *Report of the World Commission on Environment and Development: Our Common Future* (1987), generally referred to as the Brundtland Report after its chair Gro Harlem Brundtland. The commission was formed in 1983 to address growing global concerns about the effects of deterioration of the environment and consumption of natural resources on economic and social activities. The report examined a broad range of topics, from population growth and sustainable development to energy consumption, food production, and international economic relationships. Unlike many previous U.N. commissions, which tended to work *in camera*, the Brundtland Commission held a series of public hearings around the world to gather expert testimony from a broad spectrum of economists, scientists, politicians, and environmentalists.

In its second chapter, the report defined sustainable development as "development that meets the needs of the present without compromising the ability of future generations to meet their own needs."[19] The report went on to state that "contained within [this definition] are two key concepts: 1) The concept of 'needs', in particular the needs of the world's poor to which overriding priority should be given; and 2) The idea that limitations imposed by state and technology on the environment's ability to meet present and future needs." Implicit in this definition is the concept of limiting growth, and thus consumption, today so that future generations will have raw materials and energy sources to utilize for development. In turn, limiting growth requires that expanding economic activity, which has been the foundation of development since Adam Smith first inquired into the *Wealth of Nations*, be slowed or even halted.[20]

The Brundtland Commission's chapter on sustainability broadly discussed the need to balance population growth with technological innovation, reorienting technology to be more energy efficient, as well as not produce negative externalities. It called for the development of environmental risk analysis for all present and future technologies to determine their effect on natural resources and human activity, as well as merging of environmental and economic concerns in public and private decision-making. The chapter concluded with a yet unattainable notion of worldwide cooperation in economic and social matters for the benefit of all, particularly underdeveloped nations, and the environment. The Brundtland Commission also gave the sustainability movement its own version of the trinity, the Diagram of Sustainability (fig 1.1).[21]

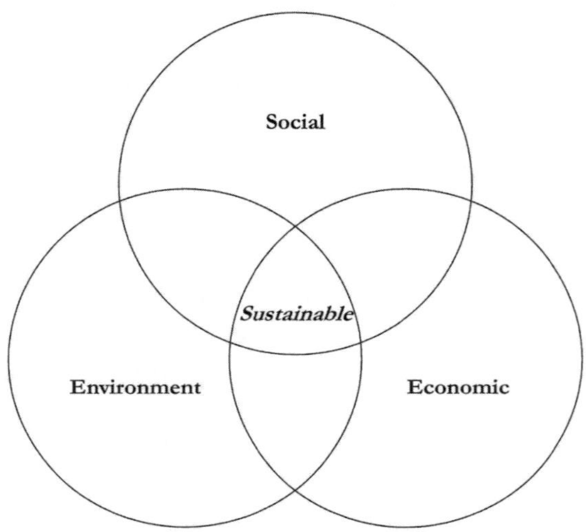

FIGURE 1.1 Diagram of Sustainability, drawn by Lea Stedt, 2010.

Not much official notice was taken of the Brundtland Report in the United States at the time of its release, perhaps because of its emphasis on transfer of wealth from richer to poorer nations. However, the seeds of a more balanced approach to development in America were planted in the 1990s with the Smart Growth movement. Embraced by many local and regional governments, the concept of concentrating population growth in existing metropolitan area has been incorporated into many state-mandated comprehensive plans. Smart Growth is based in the various conservation and environmental movements developed in America and abroad during the post–World War II period. With the rise of New Urbanism in the 1990s, with its emphasis on compact centers of human habitation based on traditional town planning and architecture, Smart Growth became increasingly popular. Planners, community development directors, and eventually local politicians began to realize that Smart Growth was a means to relieve traffic congestion since concentrated population centers could more economically be served by public transportation. It could also potentially address urban decay by re-populating urban cores. Public utilities saw the concept of Smart Growth as a means of slowing the costly extension of power lines into previously undeveloped areas, focusing expansion of services in areas they already served, thus increasing profits. With its emphasis on mixed-use development, urban theorists saw Smart Growth as a means to economic and social parity by locating affordable and market-rate housing in the same neighborhoods, with services targeting all inhabitants. Smart Growth also attracted the support of those concerned with a healthier lifestyle, where walking or biking to work replaced sitting in automobiles on the freeway during rush hour.

Thus Smart Growth had something for everyone. Over the last two decades, local planning departments adjusted or replaced zoning ordinances to allow denser development, non-Euclidian mixture of uses, and fewer parking places for automobiles.[22] It helped spawn Form Based Codes and other ordinances that paid attention to the visual environment as well as the usual zoning issues of building height, setback, floor area ratios, use, and the like.

Smart Growth is not without its critics, however. In 2001, Randal O'Toole of the Thoreau Institute published an article, "The Folly of 'Smart Growth'," suggesting that Oregon's anti-sprawl strategies have actually worsened the problems they were designed to address.[23] In 2002, Wendell Cox, a staff member of the conservative Heritage Foundation, spoke on the failure of Smart Growth to achieve its goals.[24] In 2009, a study by the University of Maryland's National Center for Smart Growth and Education found that Maryland's much-applauded Smart Growth initiative, enacted under Governor Paris Glendening a decade before, had done little if anything to slow the transformation of rural land to residential suburbs.[25]

Preservationists Look for a Seat at the Table

Historically, leadership in the sustainability movement has been within the conservation and scientific communities. They were the first to worry about the effects of humanity on natural areas, the unintended externalities of technological progress, and the political issues that must be addressed to begin to correct this imbalance. While preservationists in the United States can trace their formal roots to nearly fifty years before the founding of the Sierra Club, they have up until the last decade had only passing concern for the issue of sustainability.[26] In the aftermath of the 1970s' energy crisis, a few publications began to address the links between maintaining and preserving historic buildings and energy efficiency. In 1978, the Technical Preservation Services Division of the National Park Service published *Preservation Briefs* 3—"Conserving Energy in Historic Buildings." In it, author Baird Smith gave voice to the idea that vintage buildings, particularly those constructed prior to World War II, use less energy than those constructed in the recent past. Smith went on to discuss the three most energy-consuming systems in buildings—heat, light, and ventilation—showing how nineteenth- and early twentieth-century buildings modified interior climate with little or no energy consumption. The document also discussed actions owners of historic buildings could take to make their buildings more energy efficient without harming their historic character.

The next year, the Advisory Council on Historic Preservation published *Assessing the Energy Conservation of Historic Preservation: Methods and Examples*, a report developed by Booz, Allen & Hamilton. In it, the notion of existing buildings having value because of the energy expended for their construction—termed embodied energy—was a significant reason why we should retain and not replace them. Based on earlier work that attempted to calculate the number of BTUs it took to extract, process, transport, and incorporate typical building materials—wood, brick, plaster, stone, and the like—into a building, the Booz, Allen & Hamilton study attempted to develop a mathematical model for understanding the amount of embodied energy in existing buildings and the additional energy it would take to demolish and replace them.

The idea of embodied energy, which has been widely adopted by the preservation community, is not without its detractors. For example, the U.S. Green Building Council is not convinced that LEED points should be awarded for energy already expended and embodied in existing buildings.[27] Similarly, economists look at past expenditures, whether money or energy, as "sunk," with little or no value to present and future costs.[28] Additionally, there is no universally accepted model for calculating embodied energy in buildings and thus estimates vary widely, which is detrimental to the concepts being universally accepted.[29]

Nevertheless, the National Trust for Historic Preservation has championed the idea of historic buildings and embodied energy since its 1981 publication, *New Energy From Old Buildings*. It has made the concept the centerpiece of its current campaign to find a seat for historic preservation, and the National Trust, at the sustainability table. The organization's last sustainability campaign, begun under the presidency of Richard Moe, has developed a number of initiatives linking historic preservation to sustainability. Beginning in the middle of the first decade of the twenty-first century, the National Trust began to promote the idea that historic and other vintage buildings were inherently sustainable due to their embodied energy and original low-energy features such as operable windows, shutters, porches, and, in the case of masonry structures, thermal mass. Later, the Trust and others expanded that notion to discuss the savings in landfill by not demolishing existing buildings, as well as saving the energy it would take to tear them down and transport the debris to dumps. Toward the decade's end, the Trust popularized the phase "the greenest building is the one already built," giving preservationists an easily understood catch phrase.[30]

In 2008, the National Trust and the Friends of the National Center for Preservation Technology and Training hosted a group of preservationists, architects, green building, and energy experts at the Pocantico Conference Center on the grounds of the historic home of the Rockefeller family.[31] The result of the meeting was the "Pocantico Proclamation on Sustainability and Historic Preservation," which sought to embed preservation into the growing national and international debate on sustainability. With a nod to the Brundtland trinity, the Proclamation was based on three overlapping imperatives. The "Climate Imperative" noted that climate change caused by human activity has accelerated and placed the environment in danger, suggesting that climate change must be addressed through the reduction of greenhouse gases. The "Economic Imperative" focused on our current economy, suggesting that it is ultimately unsustainable due to an overreliance on finite resources and that it must be replaced with a "green" economy. The "Equity Imperative" noted that economic inequalities in the United States and aboard have grown to "unacceptable" levels through the consumption of resources. These must be addressed through redistribution and social and economic equity. In addition, the Pocantico Proclamation contained five principles designed to address the imperatives. These range from fostering a culture of reuse and valuing heritage, to realigning preservation policies with sustainability.

As historic preservation attempts to gain a place at the sustainable table, there are a number of fundamental issues that must be resolved. For example, as alluded to in the Pocantico Proclamation, preservationists may need to reassess some of their fundamental principles. Preservation in the United States draws its philosophical underpinnings from mid-nineteenth-century art critics and architects, most notably John Ruskin and E.-E. Viollet-le-Duc. Codified in the mid-1960s in legislation that vested the federal government's interest in preservation in the National Park Service through the creation of the National Register, state historic preservation offices, and, a decade later, the Secretary of the Interior's Standards, historic preservation was originally concerned with properties and districts of "architectural" and/or "historical" importance. Since the 1960s, the spectrum of places so deemed significant has expanded enormously to encompass such places as transportation networks, rural landscapes, traditional African American enclaves, sites associated with the Cold War, and mid-twentieth-century residential subdivisions.

Throughout the broadening process, a guiding principle of preservation has remained to minimize change to the original configuration and exterior materials. Yet as more and more inherently energy-inefficient buildings of the mid-twentieth century become eligible for listing on the National Register, the concepts of preserve rather than restore, and restore rather than replace may no longer be relevant. Consider, for example, the recent removal and total replacement of the exterior of the Lever House in New York (1951–52), an icon of corporate modernism designed by Gordon Bunshaft of Skidmore, Owings & Merrill. Some forty years later, the exterior glass curtain wall was detaching from the structure. Since it was technically impossible to save the exterior in situ, and the demand was intense to lower energy loss through the original single-pane glass, the decision was made to replace the entire exterior with a thermally improved duplicate. Not only did this significantly reduce energy consumption, it allowed the building to lower its carbon footprint. Even with this complete exterior transformation, the building retains its historic designation. Imagine the replacement of the entire exterior of a listed masonry pile such as The Rookery in Chicago (1885–88). One can readily predict the protests from preservationists and its potential de-listing as historic if such action was undertaken. However, as more and more glass curtain-wall buildings are added to the National Register, actions similar to those undertaken on the Lever House will become increasingly common.

Accommodating preservation or any other economic enterprise within a green or slow-growth economy enters equally uncharted waters. As Donovan Rypkema has written on extensively, preservation contributes to local, state, regional, and national economies through job creation, leveraging private monies for tax credits, heritage tourism, and a host of other economic activities associated with the preservation of buildings, districts, regions, and cultures.[32] All of this economic activity has occurred under the current capitalist system, where growth is good and more growth is better.

On the other hand, many eco-economists suggest that shifting to a slow or no-growth economy or a "steady-state economy" such as Herman Daly has championed, not only challenges preservationists, but society in general.[33] To achieve a steady state economically would require stabilizing the depletion of natural resources below the point at which they become exhausted. Thus before current energy sources such as oil or coal are depleted we would need to stop consuming them even if alternative sources of energy to maintain current quality of life in developed countries or improve the quality of life in underdeveloped nations were not available. Second, achieving a steady-state economy would mean the transfer of quality of life from advanced to non-advanced countries to limit inequalities between societies as called for in the Bruntland Report. Third, to maintain an ecologically based and balanced global economy would entail some means to control or even reduce the world's population, such as some form or transferable licensee to give birth. Clearly, while the notion of a steady-state economy has certain merits, it also contains the seeds of unintended externalities that contradict the moral compass of many of the world's people.

The need for a sustainable planet is undeniable. We cannot simply continue to expand our populations, consuming more and more resources, without ultimately doing irreparable harm to ourselves and the institutions and artifacts we have created. The trinity of culture, environment, and economics as first articulated in the Bruntland Report is useful shorthand for visualizing the intertwined areas that must be addressed, but reveals little if anything about the complexity of the issues involved. While it is appropriate for preservationists to become concerned with sustainability, they must understand that the topic is far more complicated than has been presented to date. Preservationists must also realize that obtaining a seat at the sustainability table may require changing some of their long-held beliefs.

Notes

1 Attributed to Quintus Septimus Florens Tertullianus (160 – 220 AD).

2 Cap-and-trade assumes that the free market will eventually produce equity between industries and countries that use large amounts of energy and those that use less, by trading in units of "carbon credits," one of which is typically defined as being equal to one metric ton of CO_2.

3 William Nordhaus is Sterling Professor of Economics, Yale University. Along with Paul Samuelson, Nordhaus is the author of the popular college textbook *Economics*, used in classrooms throughout the United States. Over the past twenty years, he has written extensively on the economics of climate change, including *Managing the Global*

Commons: The Economics of Climate Change (1994) and, with Joseph Boyer, *Warming the World: Economic Models of Global Warming* (2003).

4 Reliance on the free market is not well accepted by many involved with sustainability issues. They believe that the issue requires government intervention, pointing to a number of cities that have abandoned their recycling programs in 2010 due to the high cost of collecting the material from homes and offices and the low price obtained from recyclers, coupled with the economic constraints of the current recession.

5 An externality in economics is a by-product of a transaction that is not fully accounted for in price. They can be positive or negative. An example of a positive externality is education that is typically undertaken to improve one's employment prospects or increases one's knowledge in a particular subject. An often-cited positive externality associated with increased levels of education is reduction in the number of crimes committed by those with higher education. On the other hand, a negative externality can be the social and economic costs of air pollution from oil- and coal-fired plants that produce electricity.

6 Pigou's *Wealth and Welfare* (1912) brought social welfare into economic analysis, suggesting that government intervention through taxes or subsidies could correct negative externalities.

7 BTU, British Thermal Unit, is a traditional measure of energy. One BTU is approximately equal to the energy it takes to heat one pound of water one degree Fahrenheit.

8 In economics, income can be earned as long as one works, while capital is eventually consumed, and thus depleted, unless renewed through income.

9 Donella H. Meadows, Dennis L. Meadows, Jørgen Randers, and William Berhens III, *The Limits to Growth*. The Club of Rome is a global think tank founded in 1968 by Auerlio Peccei, an Italian industrialist, and Alexander King, a Scottish chemist, with the mission "to act as a global catalyst for change through the identification and analysis of crucial problems facing humanity." For further information, see *http://www.clubofrome.org*.

10 In *An Essay on the Principle of Population* (1798), Malthus argued that the exponential increase in the world's population would outstrip the food supply leading to society's collapse.

11 For more information on the Friends of the Earth see *http://www.foe.org*.

12 In 1955 the Pennsylvania College of Women became Chatham College and in 2007, the name was changed to Chatham University. In 1940 the U. S. Bureau of Fisheries became the U. S. Fish and Wildlife Service.

13 For more information on the founding of the British National Trust see *http://www.nationaltrust.org.uk*.

14 Unfenced (enclosed) common lands were typically used for grazing animals owned by different families. In the mid-nineteenth century, there was a movement in Britain, with parallels in the American West, to divide and fence the common lands into individual owned properties.

15 The British National Trust covers England, Wales, and Northern Ireland. Scotland's National Trust, founded in 1935, is a separate organization concerned with many of the same issues as its counterpart.

16 In contrast, the National Trust for Historic Preservation, its American counterpart, which was founded in the late 1940s, focuses on the built environment. Currently it has less than 300,000 members.

17 For further information on the Sierra Club see *http://www.sierraclub.org*.

18 The Sierra Club now has 1.3 million members.

19 Brundtland Report, Chapter 2: "Toward Sustainable Development."

20 First published in 1776, the same year as the Declaration of Independence, *An Inquiry Into the Nature and Causes of the Wealth of Nations* is probably the most influential, but least read, book on economic theory. (At almost 1200 pages, it is quite dense.) In it, Smith laid the groundwork for the capitalist system, introduced the concepts of the invisible hand of the market to regulate price and supply of goods and services and free trade, as well as the need for government to regulate the excesses of both for the benefit of all.

21 Known as a Venn diagram after its inventor John Venn, it is used to teach elementary set theory, typically showing the logical relationships between a finite set of things. Often a Venn diagram is shown as three overlapping circles with the intersections pinpointing relationships between the individual sets.

22 Euclidian zoning is named after the City of Euclid, Ohio, the defendant in a case that was eventually settled by the United States Supreme Court that allows local governments to regulate the use of land for the welfare of its citizens.

23 Randell O'Toole, "The Folly of 'Smart Growth'," *Regulation* 24 (fall 2001): 20–25. The Thoreau Institute is a West Coast organization that advocates environmental protection through incentives, not regulations.

24 Wendell Cox, "Dangers of Smart Growth Planning," testimony before the Senate Committee on Environment and Public Works, 15 May 2002.

25 It is ironic that Governor Glendening founded the National Center after he left office in 2000. Part of the reason for the failure of the statewide policy was political. Glendening's successor eliminated the Maryland Office of Smart Growth. The decision was partly based on economics; it was simply less expensive to continue to build on raw land than within existing population centers.

26 Organized historic preservation in the United States is generally believed to have begun with the founding of the Mount Vernon Ladies' Association by Ann Pamela Cunningham in 1854. The group was established to preserve the home of our first president, as well as (somewhat naively) an attempt to prevent the growing conflict between the north and south from splintering the Union.

27 The U.S. Green Building Council is a private non-profit organization that promotes Leadership in Energy and Environmental Design (LEED) as well as trains and certifies individuals to conduct LEED evaluations of buildings and awards four levels of certificates for energy conservation.

28 Sunk costs are considered retrospective costs that cannot be recovered, which are often compared to prospective (future) costs, which are those that are relevant to investment decisions.

29 For example, see Robert Costanza, "Embodied Energy and Economic Valuation," *Science* 210 (12 December 1980): 1219–24.

30 The author understands that this phase was coined by architect Carl Elefante.

31 This was not the first conference on sustainability to meet at Pocantico. In 2004, the Pew Charitable Trust hosted a two-day event on climate change at the Conference Center.

32 While Rypkema's writings on preservation and economics, which span a quarter of a century or more, are too numerous to list here, the author calls attention to

"Sustainability and Historic Preservation" given at the Historic Districts Annual Conference, New York City, 10 March, 2007; see *http://wrvation.org/rypkema.htm*.

33 Daly and economists with the World Bank and later with the University of Maryland have written extensively on the need for the world to move from its current growth economy to one with a steady state of no or very slow growth. See, for example, *Steady State Economy*, 1973; 2nd ed. (Washington, D.C.: Island Press, 1997).

Preservation, Sustainability, and Environmental Impact Assessment in Post-Colonial Developing Nations

Thomas F. King

Some years ago, I was on Pohnpei in the Caroline Islands of the western Pacific, consulting about historic preservation with the then-forming government of what would become the Federated States of Micronesia (FSM). In preparation for the FSM's independence, the United States government was funding construction of a road around the island's perimeter. One day the FSM's historic preservation staff and I were called out to resolve an unexpected conflict. The bulldozer driver clearing the right-of-way had overturned a boulder on the edge of a stream. He had been immediately set upon by the people of the nearby village, who seemed quite prepared to string him up from the nearest breadfruit tree. When we arrived, he was more or less cowering atop his rig while a crowd of irate citizens berated him. Why? Because in the eyes of the villagers the rock was an ancient, ancestral adulteress, turned to stone for her offense to the cultural norms of the people. It was not, however, on the FSM's or Pohnpei's lists of historic places; it was not known to the historic preservation office; and it was not identified in the archaeological survey that had been done to comply with both U.S. and FSM laws during an environmental impact assessment of the road project.

One might contend that the historic preservation office should have known about the rock, that the government should have had it on its register, and that the survey should have identified it—all perhaps true enough, but they did not, and that did not stop the villagers from throwing themselves at the bulldozer. The point is that places unknown to government, unregistered by officialdom, of little interest to archaeologists and architectural historians, can impede projects and poison the relationship between project proponents and the people a project may be designed to serve. And it is not only places that can cause such rifts. So can culturally valued animals, plants, water, and atmospherics. So can spatial relationships as expressed in concepts like feng shui. So can traditional beliefs and practices. Some nations—Japan, for instance—try to include such more or less non-place values in their national lists or registers of historic "places," but most do not.[1]

The road on Pohnpei was rerouted, the rock was put back in place, and the matter was (as far as I know) resolved. But it cost money, it took time, and to some extent it doubtless alienated the villagers from the project proponents and the government in general. This is a problem for any kind of development, but it is especially so for development that is supposed to be sustainable; that is, development that "meets the needs of the present without compromising the ability of future generations to meet their own needs."[2] It is hardly consistent for a sustainable development to be perceived by local people, who very likely are among those it is supposed to serve, as running roughshod over the cultural values that they wish to maintain.

So how is a sustainable development planner—say, someone working in a developing nation on a sustainable agriculture or forestry scheme, or planning a road to help small-scale farmers bring their goods to market, to ensure he or she is not going to run afoul of a rock in the stream? Or what about in our own country? A bend in northern California's Klamath River, known as Katimin by the Karuk Tribe, for example, is traditionally that people's center of the world, a place of supreme spiritual and cultural significance. Yet it is on no official register of historic places and would not be recognized for anything other than its natural attributes by anyone not informed by consulting with the Karuk people (fig. 2.1). There is a basic, but flawed, assumption on the part of planners that they can stay out of trouble with matters of cultural heritage by following the historic preservation rules of the relevant nation or funding agency.

The Encyclopedia of Life Support Systems and Historic Preservation

To help sustainable development planners in their work, around the beginning of the twenty-first century UNESCO began putting together an on-line

FIGURE 2.1. "Katimin" bend, Klamath River, California. Photograph by Thomas F. King, 2004.

"Encyclopedia of Life Support Systems" (EOLSS). I was asked to write a piece for the EOLSS on historic preservation laws.[3] The objective was to give planners an idea of what historic preservation legal requirements might pertain to projects they were planning in different countries—particularly developing countries where the bulk of the world's population lives—and how they might be most efficiently addressed. Luckily, my co-author and I were given a pretty free hand. It may be that the EOLSS's editors expected a country-by-country recitation of historic monument laws and antiquities regulations, but we quickly determined that this conventional approach would not be helpful; it would in fact be counterproductive.

As a result, our core recommendation was not to ignore the historic preservation laws of the world's nations and the guidelines and protocols of its international organizations, but not to rely on them either. We recommended regarding such laws and the institutions they create as sources of information, but not as authoritative ones. Indeed, we suggested that they are usually quite limited and possibly misleading indicators of a development project's implications for the cultural environment and of that environment for the sustainability of proposed projects. We concluded that:

> Culture is a powerful motivator, and cultural value is often ascribed to places in the landscape, whether law specifically protects such places or not. The graves and living sites of a people's ancestors, which archaeologists think of as archaeological sites, are among such places. So are natural and built landmarks thought to have spiritual significance, traditional towns and villages, landscapes in which traditional events are thought to have occurred, and a wide range of other "properties" or "places." A project that preserves such places is likely to be more acceptable to a community than is one that destroys them, whether the places involved are officially identified and given legal protection or not. As a result, it is wise for development planners to integrate programs of historic preservation appropriate to the culture of affected communities into their development plans, rather than to focus narrowly on the specific requirements of law.

What's Wrong with Historic Preservation Laws?

The central weakness that characterizes historic preservation laws—in every nation, I believe, but especially in developing nations—springs from their grounding in European common law. Typically, laws thus based provide for the formal recognition of historic places, usually through inclusion in some kind of list, register, or schedule. The National Register of Historic Places in the United States is an example of such a list, as is the World Heritage List in an international context. Some laws essentially stop at this point; they are vague about what rights and responsibilities are conveyed by such registration to the owner or user of a designated property. Others go on to apply particular strictures, sometimes going so far as to define a scheduled property as inviolate—dictating that it cannot be destroyed. Many modern laws, though, provide for the consideration of designated properties in planning without insisting on their physical protection in perpetuity. This leniency creates an obvious interface with development. In planning a project, it seems only reasonable to secure the nation's official listing of historic places and address any listed properties in the proposed scheme.

There are several problems with this approach that can get planners in trouble and complicate, indeed thwart, the interests of sustainability. First, the very fact that the system of registration and protection is derived from European law means that it was imposed on or adopted by other countries in some sort of colonial context. It is not an outgrowth of the cultures of most nations. Thus a legal system for treating cultural heritage may be foreign to that heritage and not something to which a nation's population easily relates. Second, the nation's list of historic places must be maintained by someone, which has led nations to set up cultural ministries, historical and archaeological agencies, and other such government entities, staffed by civil servants and sometimes academics, charged with keeping the list and ruling on whether something newly recognized—a freshly discovered archaeological site, for instance—should be entered on it. To make such judgments, the office must establish criteria and procedures, creating bureaucratic structures that may become quite impenetrable by planners and the public alike.

A third problem, familiar to preservation practitioners in the United States and Europe, is that a nation's list is invariably, necessarily, incomplete. It does not include everything in the nation that has cultural and historical significance simply because no one has found everything or invested the effort necessary to register everything found. Different countries have dealt with this problem in different ways. Some have effectively ignored it. Others have charged their cultural ministries with conducting surveys to "complete" their lists. Yet others have charged such agencies with conducting quick surveys in advance of development projects to feed information on hitherto unlisted or unknown places into project planning. And some, including the United States, have charged the project's proponent with the responsibility to find, document, and address such places. Each system has its strengths and weaknesses, but they all mean that a development planner cannot assume that he or she can just walk into the government's cultural agency and check the list or call it up on the World Wide Web. But the very existence of a national list, and government's tendency to describe it in grandiloquent terms, can lead a planner to assume that this is precisely what he or she can do. For instance, the U.S. National Register describes itself as "the Nation's official list of cultural resources worthy of preservation."[4] If one does not know better, this description suggests that one need only check the register to find out what is "worthy."

Finally, the experts in historic preservation offices naturally apply their professional judgment in making their listing decisions. This may be fine as far as it goes, but there is not necessarily much interface between professional standards and the cultural values that a local population wants to sustain. A place that seems very significant to an archaeologist or architectural historian may induce only yawns from the citizenry. At least equally problematical in terms of sustainable development, a place that citizens regard as tremendously significant to their cultural integrity—a rock in the stream—may be utterly unrecognizable and unvalued by professionals.

The result of all these factors is that the historic preservation laws of many nations do not ascribe significance to the kinds of places that the nation's people and communities value as parts of their cultural heritage. The laws tempt development planners to focus their attention on official lists of places that a professionally grounded governmental system has defined as significant—lists that almost invariably are incomplete and biased. The laws further involve the application of complicated systems of regulation, criteria, and standards that may be sensible only to the government employees who administer them. Finally, the laws may be regarded by government and the public alike as irrelevant and even offensive holdovers from the days of colonialism. Recognizing all this, we recommended a different approach.

Environmental Impact Assessment and Sustainability

In 1969, the U.S. Congress enacted the National Environmental Policy Act (NEPA). Over the next several decades, many if not most other nations in the world put more or less similar laws into place and international organizations like the World Bank adopted similar policies. Collectively the various laws, regulations, and policies establish programs of environmental impact assessment (EIA). EIA involves examining the potential environmental impact of proposed actions and alternatives to such actions. What will happen to the environment if one builds this dam, this highway, this housing project? What would happen if one pursues this alternative to the dam, the highway, the housing project, to achieve roughly the same purpose? What about this other alternative; what effect would it have? Typically, an EIA law does not forbid unsustainable development, nor does it forbid messing up the environment, but it does require, in theory at least, that governments look before they leap, think about what damage their proposed action may do before they actually do it, and what alternatives may produce a less damaging, more sustainable result.

EIA, if it is properly done—and I hasten to say that it often is not—can be critically important to promoting sustainability. As part of an assessment we can ask ourselves which alternative way to get water or power, move people from here to there, or house a population will be most sustainable and factor the answer into our decision-making about whether and how to meet the defined need. At the same time, sustainable development schemes cannot be insulated from the scrutiny that EIA affords. Even the best schemes, the ones with the most long-lasting beneficial effects, may still do damage. If it is decided to move forward with a good but damaging plan, actions ought to be taken to mitigate whatever damage it will do. One obviously cannot deal with the damage if one does not assess what it is.

One aspect of the environment that is typically examined in an EIA comprises the environment's cultural aspects, which are called things such as heritage, cultural resources, or historic and archaeological properties. All these terms present the EIA practitioner and sustainable development planner with some problems.

The World Bank Standard

The World Bank, which funds a lot of sustainable development and, unfortunately, some that is not so sustainable, requires an EIA on projects it supports. The bank's guidance on the matter states that "when relevant," an environmental assessment on a development project must address impacts on "archaeological sites, historic monuments, and historic settlements," as well as the "traditional land rights" of indigenous peoples.[5] This requirement highlights the need to plan development projects, whether sustainable or not, with sensitivity toward protecting culturally valued aspects of the land. The devilish word in this guidance is "relevant." When is it "relevant" to consider impacts on historic and cultural places? The danger inherent in the historic preservation laws of many nations is that such "relevance" is likely to be recognized only if the project is going to affect something on the official listing of historic places. Or perhaps "relevance" will be recognized when a governmental, cultural, or historical bureau says it exists. Or historic and cultural places may be examined in EIA, but only by applying the nation's official criteria of historical significance to places found through professional study. All these approaches can lead project planners into "rock-in-the-stream" types of conflict.

But in many aspects of EIA—considering the sensitivity of an ecosystem, effects on the quality of the water, impacts on local economics—one does not have an official list, a resource-specific government bureaucracy, or detailed governmental-academic criteria to apply. One looks at the environment and applies one's judgment, based on what one may hope are sound analytic standards. This is what we suggested that sustainable development planners do with regard to historic preservation. Do not ignore the preservation laws, criteria, and standards, but do not take them too seriously.

We proceeded to outline some commonsense guidelines for historic preservation in EIA on sustainable development projects:

- Identify historic places—and other culturally valued aspects of the environment like traditional ways of life and culturally significant animals or plants—using whatever procedures make sense given the character and history of the potentially affected area and the kinds of changes likely to result from the project. This will almost always involve some sort of original research; registers and other official information sources provide one useful set of inputs to identification, but only one. Importantly, it will always involve consultation with the people of the area, seeking out parts of the environment that are important to them as well as those of interest to science, tourism, and other external concerns. If some such places cannot be identified precisely because of cultural sensitivities, they can be identified more generally, to the extent necessary to manage impacts on them.

- Identification necessarily includes evaluation—figuring out which places and things are worth worrying about and which are not—but do not evaluate solely in terms of criteria for putting something on the government's list. Generally, if someone thinks something is important—be that someone the members of a local community, an academic body, or the cultural ministry—it is important enough to consider in EIA. Focusing only on what is important to government, or to a post-colonial elite, is a formula both for unsustainable development and for trouble. So evaluation, too, has to be based on consultation with local people, as well as with government and academia.
- Integrate the valued aspects of the cultural environment into project planning—if possible in a positive manner. Planning for such integration also demands consultation with the people who ascribe value to aspects of the environment. Does the community want tourists to visit its spirit mountain, or not? Is traditional farm technology important to maintaining the community's family structure, or can we employ more modern tools? A plan for such integration will probably need to be developed, in consultation with affected and concerned people, and probably approved by government. Where a project requires the destruction or alteration of a culturally significant place—and this is often the case, even with a sustainable development project—a plan should be developed and carried out to preserve the place as much as feasible, or recover relevant information from it in a manner that's acceptable to local communities.

These steps may seem obvious, but many national historic preservation laws in effect discourage such culturally sensitive planning.

Akwé: Kon

Thoughtful, creative, culturally sensitive, and timely EIA is one of the keys to making sustainable development work. Historic preservationists ought to be trying hard to make our national and international systems work to facilitate this kind of EIA. Unfortunately, I see little evidence that historic preservation practitioners are much interested in rethinking the often counterproductive systems we have built. Instead we continue to split academic hairs over what is eligible for a given historical register, while governments seek to "streamline" their EIA procedures by limiting and overstructuring public involvement. But if the historic preservation community is flaccid in its attention to EIA and befuddled by its traditional reliance on national schedules and registers, others are not, and in this latter direction there may be hope. I was astonished and gratified recently to learn that in 2003 the Secretariat of the Convention on Biological Diversity issued guidelines for EIA that provide direction very similar to what we offered, but did so in much greater detail. These guidelines, called Akwé: Kon and pronounced "agwe goo," are succinctly subtitled "voluntary guidelines for the conduct of cultural, environmental and social impact assessments regarding developments proposed to take place on, or which are likely to impact on, sacred sites and on lands and waters traditionally occupied or used by indigenous or local communities."[6] Appropriately enough, akwé: kon is a Mohawk term meaning "everything in creation."

The Akwé: Kon guidelines lay out recommended actions in logical sequence. Early in developing any project, planners should begin notifying and consulting potentially affected communities, and through this consultation should identify stakeholders to consult further. Very importantly, they should establish "effective mechanisms for indigenous and local community participation, including for the participation of women, the young, the elderly and other vulnerable groups, in the impact assessment processes." Do not expect local people, indigenous people, old and young people to file written comments or compress all their concerns into five-minute statements at public hearings; design a process that includes them in a manner sensitive to their modes of communication and cultural values. It is also important to establish a process that is agreed upon among the stakeholders for recording the views of the people and groups that participate.

The guidelines further emphasize the need to establish a process "whereby local and indigenous communities may have the option to accept or oppose a proposed development that may impact on their community." This stipulation may seem obvious, but a local group may not understand that it can do anything but acquiesce in a government proposal, and the planning process may tend to validate this misunderstanding. A community may or may not actually have the ability to stop a project, but it should certainly have a meaningful way to oppose it and know that it can do so. There is further the need to identify "sufficient human, financial, technical and legal resources for effective indigenous and local community participation in all phases of impact assessment procedures." Expect and prepare to cover the costs of consultation. Through consultation with the stakeholders—assuming the project goes forward—establish a plan for managing and monitoring it to make sure that impacts are dealt with, and establish who is responsible for "liability, redress, insurance and compensation" if something goes wrong. Agreements should be concluded with the stakeholders about how the plans will be carried out, and then implemented, providing for review and appeals if there are disputes. Those are the procedural aspects of the guidelines. Their substantive aspect is a recommendation that EIA should include cultural and social as well as more traditionally environmental impact analyses.

A cultural impact assessment, say the guidelines, should seek to identify "the issues that are of particular cultural concern ... such as cultural heritage, religions, beliefs and sacred teachings, customary practices, forms of social organization, systems of natural resource use, including patterns of land use, places of cultural significance, economic valuation of cultural resources, sacred sites, ceremonies, languages, customary law systems, and political structures, roles and customs." Not just "historic places," though such places should certainly be identified in the process of such a holistic examination, with or without reference to national registers and schedules.

The cultural impact assessment should go on to analyze the project's potential impacts on such things as customary uses of natural resources, traditional knowledge and practices, secret and sacred knowledge (following protocols to ensure that it stays secret if need be), customary law, and both "sacred sites" and the activities associated with them. Assessments of impacts on the natural environment should also take cultural factors into account, looking at such things as the species important to the community for food, medicine, fuel, fodder, construction, spiritual activities, the habitats of culturally and economically sensitive animals, places in the landscape that have spiritual significance to the community.

Social impact assessments should evaluate how the project may change traditional "gender and demographic factors, housing and accommodation, employment, infrastructure and services, income and asset distribution, traditional systems and means of production, as well as educational needs, technical skills and financial implications." The guidelines draw special attention to the potential for effects on traditional ways of life, land tenure and land use, and social cohesion. All these kinds of impact assessments are supposed to be integrated to achieve the end of finding out what is important about the area to be affected from, among others, a cultural point of view, and figuring out what to do to manage change in a way that does minimal harm and maximum good from that point of view. And all of this is supposed to be done in respectful consultation with the people affected by the action.

Without belaboring the matter further, it seems to me that if a nation adopted the Akwé: Kon guidelines as a core standard for the conduct of EIA, that nation's EIA practitioners and sustainable development planners could happily forget about national registers, schedules, lists, and historic preservation regulations and take excellent care of that country's heritage. Yes, they would doubtless need to make sure that archaeologists and architectural historians were represented on the assessment team or that their views were sought along with those of the local communities. But the central purpose of EIA's consideration of the cultural environment ought to be to identify and manage that cultural environment in a manner that is most satisfactory to those who use and value it. The purpose should not be to identify and manage places that qualify for some bureaucratic government list, or to satisfy esoteric legal standards inherited from the nation's onetime colonial overseers.

Notes

1 UNESCO's "Convention on Safeguarding of the Intangible Cultural Heritage" recommends such inclusion; see *http://www.unesco.org/culture/ich/index.php*. A national or international list of officially sanctioned intangible cultural heritage strikes me as an even worse idea than such a list of places and other tangible things.

2 "Report of the World Commission on Environment and Development," United Nations General Assembly Resolution 42/187, 11 December 1987. See *http://un-documents.net/wced-ocf.htm*, accessed 2 August 2008.

3 See *http://www.eolss.net/*. My contribution was done with the assistance of Sam Struelsen, then a graduate student at George Mason University.

4 See *http://www.nationalregisterofhistoricplaces.com/*.

5 "Policies, Procedures, and Operational Directions for Environmental Assessment," World Bank, 2000; see *http://wbln0018.worldbank.org/essd/essd.nsf/EnvironmentalAssessment/Overview*.

6 "Akwé: Kon, Voluntary Guidelines for the Conduct of Cultural, Environmental and Social Impact Assessments Regarding Developments Proposed to Take Place On, or Which Are Likely to Impact On, Sacred Sites and on Lands and Waters Traditionally Occupied or Used by Indigenous and Local Communities," Secretariat of the Convention on Biological Diversity, Montreal, 2004; see *http://www.cbd.int/doc/publications/akwe-brochure-en.pdf*.

The Massachusetts Heritage Landscape Inventory: A Community-Based Approach to Identifying and Protecting Special Places

Shary Page Berg

Around the country citizens are coming together to create a vision of what their community might be and to develop steps toward making these visions come true. Alternatively called "healthy," livable" or sustainable communities, these efforts are integrative, inclusive and participatory. In many communities—large and small, rural and urban—issues are being addressed in an interconnected manner. They are demonstrating how innovative strategies can produce communities that are more environmentally sound, economically prosperous, and socially equitable. Sustainable Communities Network [1]

Clearly the concept of sustainability, which was initially identified primarily with buildings, can also be applied to communities—to preserving neighborhoods, to accommodating growth in existing communities, and to developing connections between preservationists and other stakeholders. On the other hand, traditional preservation techniques such as historic resource surveys, preservation plans, and listing on the National Register of Historic Places are powerful tools for dealing with buildings, but are often less effective in dealing with other resource types and with other values that contribute to preserving the quality of life in a community or region. This paper explores one program that has been particularly successful in bringing awareness to a wide range of landscape resources in communities throughout Massachusetts.

Massachusetts Heritage Landscape Inventory Program

The Massachusetts Department of Conservation and Recreation (DCR) has a long history of innovative landscape preservation programs. These include an early statewide scenic landscape inventory; the Olmsted Historic Landscape Preservation Program; the landscape preservation grant program; thematic efforts such as cemetery, town commons, and parkways initiatives; and a variety of technical assistance publications. The Heritage Landscape Inventory Program was established by DCR in 2001. The goals of the program are to increase awareness of the broad range of unprotected landscapes that have both natural and cultural values and to encourage communities, state agencies and non-profit organizations to collaborate on conservation and preservation issues. The program was an initiative of the DCR rather than Massachusetts Historical Commission (MHC) because of DCR's familiarity with landscape resources.

A key premise of the program is to expand the parameters of historic surveys to encompass neglected resources, especially those that have not been addressed in traditional surveys, that are valued by the community, and that are potentially threatened. For the purpose of the program, heritage landscapes are broadly defined as "special places created by human interaction with the natural environment that help define the character of a community and reflect its past." The program has five major goals: 1) to identify the unique historic, scenic, and environmental qualities that define each community, especially undocumented resources that fall outside the typical purview of historic preservation or open space inventories; 2) to bring together citizens and officials from various sectors of a community, whose involvement in municipal affairs is often fragmented and focused on regulatory concerns; 3) to provide technical assistance to communities, many of which lack staff or funding to accomplish inventory work or to implement sophisticated planning techniques; 4) to identify threatened resources of regional significance and to build broad constituencies for their preservation; and 5) to identify statewide issues and needs regarding specific land parcels, resource types, and land-use and preservation policies.

Over the past nine years, 108 communities, covering over one-third of the state, and eight regional partners have participated in the program, identifying more than 5,000 landscapes in a wide range of resource types. Participating communities have included cities, suburban towns, and small villages representing a wide range of economic situations and land-use patterns. Small communities with little or no municipal staff have particularly welcomed the expertise that the program brings. While the program has primarily operated on a regional basis through DCR, one suburban community has undertaken an expanded and customized version of the Historic Landscape Inventory Program on its own, using consultants with previous experience in the program (fig. 3.1).

The Heritage Landscape Inventory Program is administered through the Office of Cultural Resources at DCR, which typically works in collaboration with a partner, such as a regional planning agency or heritage area commission. Consultants with backgrounds in historic preservation, planning, archaeology, and related disciplines are hired to work in each region. Communities participate at no cost but are asked to meet certain requirements, such as appointing a local program coordinator and organizing community meetings. A cornerstone of the program is its handbook *Reading the Land*, which received the Public Education Award from the American Planning

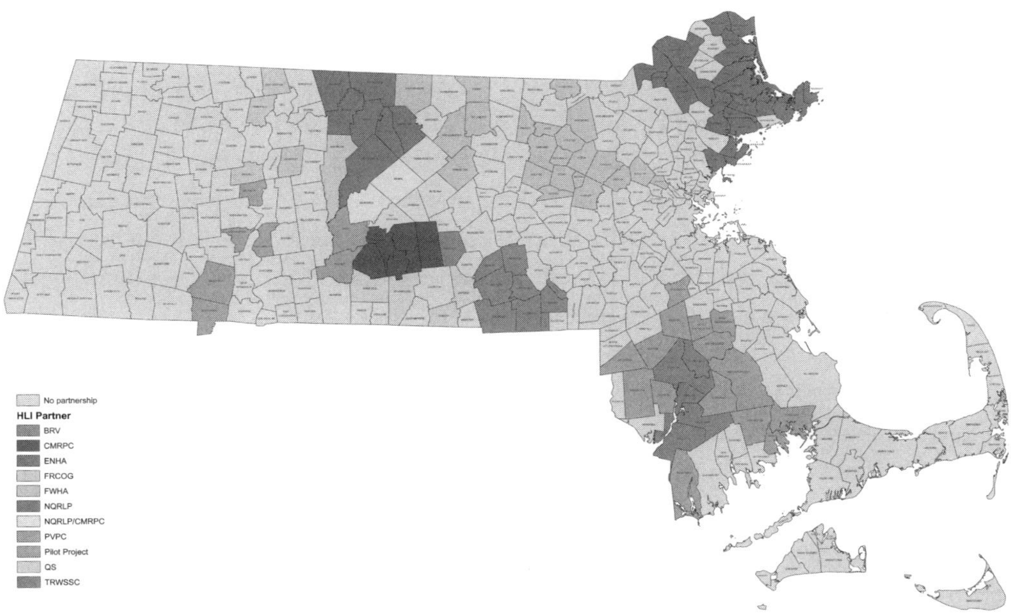

FIGURE 3.1 Map showing communities that have participated in the Massachusetts Heritage Landscape Inventory Program through 2009. (Courtesy of the Massachusetts Department of Conservation and Recreation GIS Office).

Association in 2004.[2] The publication has been widely distributed throughout the United States and abroad.

In each community, the process begins with a meeting involving participants from multiple groups and constituencies: officials from various parts of municipal government, non-profit organizations, business leaders, and other citizens. Some communities have had schoolchildren participate as part of their social studies curriculum. One community established a blog for residents who were unable to attend the meeting. At the meeting there is a brief presentation about the program and then participants are encouraged to think broadly about the values that define the character of their community and to identify special places, especially those that are not well-known. The vast majority of participants have grasped the concept of heritage landscapes and have eagerly applied it to their community. Brainstorming sessions, which are recorded by the consultants, are almost always lively and enthusiastic. One concept that caught the imagination of many people was "places of the heart," a catchall category that was used for everything from favorite sledding hills to parade routes to a favorite ice cream shop. One community held a heritage landscape photography contest to promote interest in the program.

Lists generated by communities typically range from thirty to one hundred entries. Landscapes may be small, such as historic trees, mile markers, or horse troughs, or they may be large, such as a river corridor, neighborhood, or significant open space. They almost always represent a range of land-use types. Next, participants are asked to review the list of landscapes identified by the community and to select five to ten of them that are most important to the community, are threatened, and are unprotected. On a follow-up visit, project consultants and a small group of community members visit the priority landscapes to learn more about them. The final product for each community is a reconnaissance report, about thirty to forty pages long, that makes recommendations for the priority landscapes; describes major preservation and planning issues identified by the community; provides a customized toolkit to address these issues; and includes a chart with the full list of heritage landscapes.

Once work has been completed in each community, a small group of regionally significant properties are identified for further study. These are typically either multi-community resources or represent outstanding properties or typical issues facing communities. Initially the focus was on documenting these high-priority resources on MHC inventory forms, but over time the emphasis shifted to planning assistance.

Landscape Categories

The best way to understand the range and diversity of the 5,000 landscapes identified through the program to date is to look at the major categories identified by communities. The four types that accounted for more than sixty percent of the total were: natural, agricultural, residential, and transportation landscapes.

Natural landscapes were the largest single category (about 17 percent). Typical nominations included rivers, lakes, and other water bodies, as well as geologic features and wetlands. While many of these had long been valued for their scenic and natural qualities, the program brought new awareness to the cultural aspects of places such as the Assonet Cedar Swamp in the town of Lakeville, a large, wooded wetland with a long history of swamp-related land uses such as gathering cranberries and harvesting wood for shingles.[3] A common theme linking the area and distinguishing it from its surroundings was an ongoing history of land-use patterns related to the natural landscape.

Agriculture was also a large category (about 15 percent). In rural areas, loss of agricultural land and the infrastructure to support it has resulted in profound economic and land-use changes. At the same time, people occupying more intensely developed areas place a high value on the remnant agricultural landscapes in their midst. The Woodsom Farm in Amesbury, an area of outstanding natural beauty that was once the largest dairy farm in Essex County, for example, was purchased by the community when agricultural advocates, recreational enthusiasts, historic preservationists, and affordable housing constituents worked together to acquire the 379-acre property to meet a number of community needs while preserving extant buildings and continuing agricultural use.

Residential use also represented about 15 percent of the landscapes identified. Neighborhoods were particularly valued in industrial cities such as Lawrence and Fall River, where it was the neighborhood rather than the community as a whole to which many individuals related. Urban residents also identified with neighborhood institutions, such as churches, schools, and parks, that contributed to a strong sense of community. Residents of several suburban communities felt that their mid-twentieth-century neighborhoods deserved recognition.

Transportation uses represented about 15 percent the total as well. Scenic roads and bicycle trails were identified in numerous cases, as were early mile markers and several historic airfields. While many communities had enacted a scenic roads bylaw, most were not based on documentation of the values to be preserved. One of the technical assistance projects was to create a prototype scenic roads inventory for the town of Harvard.

Cemeteries and burial grounds were another category identified by many communities, whose residents valued their scenic and historic qualities but often lacked the funds and expertise to protect them. Even before the inventory project got underway, DCR established the Historic Cemeteries Initiative, which provided technical assistance to a number of communities throughout the commonwealth in the form of "mini-master plans" and created a widely distributed handbook with preservation guidelines.[4]

Open space and recreational landscapes were identified in every community. They included public parks, wildlife refuges, reservations, natural areas, golf courses, and private property such as estates or farms that no longer continued in agricultural use. Open-space issues typically revolved around ownership, funding, vulnerability to change, and access to certain areas. Stewardship of open space and inadequate maintenance funding were also concerns, particularly in urban areas.

Industrial properties were also surprisingly well represented. They included many types of mills, often in association with the waterways that powered them. Some were still active; others had been converted to new uses such as housing; still others were in ruins. Many of the mill sites reflected potential conflicts between historic and natural resources, such as a failing dam that was blown up a few days after our visit because it posed a flooding risk to communities downstream. In several other cases a fish ladder had been installed adjacent to a dam as part of an effort to re-establish a viable fishery. Quarries, many of them inactive, were identified in several coastal communities.

Institutional landscapes included large, prominent properties, such as state hospitals and religious properties that had been closed and were facing redevelopment, as well as college campuses that were facing either closure or expansion. Civic properties included several town halls threatened by demolition, such as the 150-year-old one in Wenham where the program assisted residents in organizing a successful preservation campaign.

A number of communities in Essex County identified coastal resources as heritage landscapes. The natural features of the estuaries, beaches, and rugged coastline helped to shape the historical development of communities along the coast. The man-made interaction with these natural features is seen in the wharves, shipyards, landings, navigational markers, and ways or paths to the water. Shorefront neighborhoods are also important features that have evolved due to their proximity to the natural landscape. The issues are complex due to the economics of coastal industries, as well as the high cost of maintaining the infrastructure, the high value that the market has imposed on waterfront property, changing natural conditions, the fragile ecological balances between nature and humans, and access to these landscapes, both physically and visually.

One of the oldest and most iconic properties identified was an ethnographic landscape, the Council Oak in Dighton, located on a bluff near the Taunton River. The tree had served as a meeting place for the Wamapanoag tribe for hundreds of years but was in decline when we visited it (fig. 3.2). A few years later it was struck by lightning and killed. The tribe has been given use of the nearby Grange Hall but sadly the tree's site now overlooks the new Council Oak Estates.

Archaeological sites did not figure prominently in the survey; among the few that were identified, most had been used for industrial purposes.

Selected Landscapes

Of all the places identified by communities around the state, two in Essex County exemplify the richness of heritage landscapes and the complexity of protecting them. The area known as the Common Pasture, located in the towns of Newbury, Newburyport, and West Newbury, is a remarkable remnant landscape with a history that dates to 1635. It is one of the few surviving areas in eastern Massachusetts to retain evidence of the common lands that were the foundation of seventeenth-century land-use patterns. The grasslands that characterize portions of the surviving area of the Common Pasture are rare as a historic landscape type in the state and are visible from multiple public ways including Interstate 95,

FIGURE 3.2 Council Oak, Dighton, Massachusetts, view taken several years before it was struck by lightning. Photograph by Shary Page Berg, 2001.

which runs through the western part of the area (fig. 3.3). Most of the land is not protected. Recently there has been strong interest in its preservation due to the incremental losses at its edges and the intense development pressures that it faces. The Common Pasture's historical significance, its value as a working agricultural landscape, its size and unique physical character, and its importance as open space, wildlife habitat, and aquifer all provide compelling reasons for protection. The extant portion of the Common Pasture is a roughly 1,000-acre area that was part of a much larger seventeenth-century commons. During the eighteenth century, the pastureland was divided into family farms and has remained largely in agricultural use ever since. It contains most of the unique grasslands and remaining working farms in the area, but is in a prime location that is also vulnerable to development.

The environmental and open-space values of the Common Pasture have long been recognized. The Massachusetts Natural Heritage and Endangered Species Program has designated the area as a priority habitat that is home to rare species such as the upland sandpiper, American bittern, northern harrier, Long's bulrush, spotted turtle, barn owl, and blue-spotted salamander. The Common Pasture also supports a wide variety of both wetland and upland plants and animals that require its unusual moist open space ecology.

The historical significance of the Common Pasture derives from its early history as common land as well as its continued agricultural use as family-

FIGURE 3.3 Common Pasture, Newbury, Newburyport, and West Newbury, Massachusetts, view looking east from Interstate Route 95. Photograph by Shary Page Berg, 2005.

owned farms from the early eighteenth century to the present. In the context of Massachusetts settlement, the Common Pasture is of statewide significance as a rare surviving area where seventeenth-century land-use patterns are still evident in the landscape. The Common Pasture is also one of the largest and most cohesive agricultural areas in Essex County, with regional significance that extends beyond the three communities in which it is located. Ironically, while Interstate 95 is an intrusion into the Common Pasture, it also is a prime location for viewing the landscape, making it visible to thousands of travelers daily. Like most rural historic districts, the Common Pasture was not frozen in time but has evolved over nearly 400 years of continuous use while retaining its essential agricultural character.

While the natural, scenic, and agricultural aspects of the Common Pasture, have been acknowledged for some time, it is only recently that its historical significance has been recognized, in large measure because of the Heritage Landscape Inventory Program. Selection of the site as a regionally significant landscape as part of the Essex County inventory led to the preparation of a detailed MHC inventory form, the first effort to fully document the historic resources within the area. This in turn led to the Common Pasture being listed in 2004 as one of the "Ten Most Endangered Historic Resources of Massachusetts" by Preservation Massachusetts, a statewide advocacy group. The citation stated in part:

> As is the case for many existing open spaces, development is a constant threat. With its extremely large expanse, the pasture is especially challenged since it is controlled by three different sets of municipal zoning regulations representing each community through which it spreads. Residents have expressed concern for the pasture's future at town meetings and encouraged town officials to work with appropriate agencies to secure formal protections for the entire Common Pasture area. The imminent threat is evidenced in that permitting processes are already underway for proposed development on a wooded section near the south end of the pasture. The land represents a feeling of "home" to the community and preserves our working heritage.[5]

Most recently, municipalities, regional groups, and national organizations such as the Trust for Public Land have been working together to protect key portions of the Common Pasture. The project involves multiple strategies including land acquisition and agricultural preservation restrictions. Agreements are currently in place to acquire some parcels and to protect fifty acres of working farmland through an agricultural preservation restriction.

A second example, Dogtown, is a 3,000-acre area with diverse cultural and natural resources located on Cape Ann northeast of Boston (fig. 3.4). Dogtown is a mysterious, complex, rugged, difficult-to-read area that is overgrown and has not been well mapped. While the terrain is rocky and undeveloped, it has a long and diverse history, which is evident in the landscape. Key historic features include the commons settlement, common pasture, historic woodlots, several reservoirs, old quarries, seventeenth-century mill ruins, and a series of inscribed boulders dating to the 1930s. Collectively these features tell a remarkable story of land use from the Colonial period until the present. Dogtown is also noteworthy for its unusual geologic and ecological features. It is located in two towns, Gloucester and Rockport, both of which identified it as a priority heritage landscape. The core of the area is owned by the two communities, but the perimeter, which retains many of the same land-use patterns, is privately held by multiple owners. Dogtown has many passionate admirers but does not have a strong, united coalition, and much of the understanding of its resources is based on outdated and fragmented information.

FIGURE 3.4 Dogtown, Gloucester, and Rockport, Massachusetts, view of 1930s boulder carving. Photograph by Shary Page Berg, 2005.

Conceptually, Dogtown can be viewed as three concentric circles. The outer circle, for which the boundaries are not well defined, encompasses about 1,400 acres of undeveloped, primarily privately held land that is vulnerable to change. This area reflects many of the same historical themes and types of resources as the 1,600 acres of publicly owned land that forms the middle circle of Dogtown. The publicly owned land has the highest concentration of historic resources and is also more accessible. The inner circle includes the remains of the commons settlement, an area of roughly one hundred acres that is a significant archaeological resource, containing forty house sites, three mill sites, and other resources associated with the Colonial period. Visible remnants of the commons settlement include the stone walls, roads, and cellar holes of the village center, as well as a more extensive system of roads and trails, common pasture land, and former woodlots associated with the early years of Dogtown. There is also a late nineteenth-century railroad line, two twentieth-century reservoirs, and a collection of quixotic, carved boulders featuring inscriptions such as "Be on Time" and "Get a Job" that were carved out by unemployed quarry workers in the 1930s.

While the MHC inventory form prepared as part of the Heritage Landscape Inventory is far from being a complete accounting of cultural resources, it is more extensive than earlier surveys. The core area of the commons settlement has already been determined eligible for listing on the National Register by the MHC as a rare surviving site of a Colonial village with a high level of integrity. Perhaps most importantly, the effort brought attention to Dogtown's needs and provided baseline data for future surveys.

Both of these examples illustrate large areas that have a wide range of natural, scenic, and historical values and are difficult to document and protect. The role of the Heritage Landscape Inventory Program in each case was threefold: to bring community awareness to the area, to document historic resources that had not previously been well understood, and to serve as a catalyst for establishing a constituency for each landscape.

Outcomes

The results of the program have been substantial at the community, regional, and state levels. Almost every community that participated in the program embraced the effort wholeheartedly and was pleased with the informal process, the user-friendly report, and the increased understanding of landscape heritage. A major benefit has been resource identification. The extensive landscape lists generated by each community have been wide-ranging and many include landscapes that have fallen through the cracks of traditional planning processes. While these landscapes are not documented in detail, their presence on the list is an important planning tool for communities. Of scarcely less importance has been prioritization. The priority landscapes selected by each community are the cornerstone of the program, as these represent local consensus of which properties are significant, threatened, and unprotected. Concise descriptions of the values of each; the issues it is facing; and recommended actions have been important tools, particularly for small communities that have not done extensive planning. Parcels identified through the program, particularly the priority landscapes, received points toward state funding programs.

A third benefit is that many communities have identified key issues—sometimes broad ones, such as threats to village centers, loss of scenic roads, poor maintenance; sometimes small ones, such as closing of a post office, church, or school. Their presence in the report validates them as community concerns. Many communities cited the benefits of coming together in a way that is proactive rather than reactive and that leads people from diverse perspectives to coalescence around shared concerns. Small communities with little or no staff were particularly appreciative of technical assistance contained in the report and obtained through DCR staff and consultants. The program often provided guidance for taking critical next steps, such as implementing a scenic roads bylaw, protecting a key parcel, or developing a preservation plan for a historic cemetery. Finally, many communities had positive outcomes that were a direct result of the program.

At the regional level the outcomes were similar to those at the community level but on a larger scale. The process brought attention to regionally significant landscapes such as Common Pasture and Dogtown, and resulted in some important coalitions and preservation success stories. The program also helped to identify regionally significant issues such as coastal resources, agriculture, and institutional properties. Regional forums and workshops have brought communities together to share joint problems and learn from each other. At the state level, the program has identified important landscape resource types and issues that pose a particular challenge for municipalities. DCR has developed a series of preservation bulletins, known as *Terra Firma*, on topics such as burial grounds, town commons, scenic roads, and stone walls.[6] The agency also provides a monthly e-bulletin to all communities that have participated in the program.

As with all public initiatives, funding is a key issue. A tool that works closely with the Heritage Landscape Inventory Program is the Community

Preservation Act of 2000, creating a program that provides funding for open space, historic preservation, and affordable housing through a tax surcharge, which is matched by the state. To date 142 communities out of 351 have adopted the act, with surcharges ranging from .5 to 3 percent of local property tax assessment.[7]

The program has brought people together to identify shared priorities for their community and region. It has forged new coalitions between town officials, non-profit organizations, business leaders, and citizens of all ages. It has also strengthened connections between communities, regional organizations, and state agencies. Despite its many accomplishments, the program has been fairly inexpensive. After an initial investment of $200,000, funding levels have been at around $100,00 per year over an eight-year period. The ninety-six communities that have already participated in the program have received the most benefit. There are also ongoing regional workshops, as well as technical assistance publications such as *Reading the Land* and the *Terra Firma* series that have reached a national audience.

The program has combined documentation, planning, and constituency building and has grappled with categories and combinations of heritage resources that have traditionally been difficult to document or understand. Of particular importance to the preservation community is the expanded understanding of the state's historic landscape resources through the 5,000 properties identified to date. As a result, the preservation community has also become a more active participant in important land preservation projects such as Dogtown and the Common Pasture. Perhaps most important is the fact that, unlike most preservation surveys, which are based on priorities established by professionals, this program is based on community values and priorities. While this type of visioning is widely used for planning projects, it has not generally been applied to historic resource inventories. If the characteristics that make a community sustainable are that it is "integrative, inclusive and participatory" and that "issues are being addressed in an interconnected manner," as described by the Sustainable Communities Network, then the Heritage Landscape Inventory Program embodies these characteristics.[8]

Notes

1 See *http://www.sustainable.org/information/aboutsuscom.html*, accessed 6 September 2008.

2 *Reading the Land—Massachusetts Heritage Landscapes: A Guide to Identification and Protection*, (Boston: Massachusetts Department of Conservation and Recreation, Executive Office of Environmental Affairs), 2003.

3 In New England and New York State "town" is the jurisdictional designation equivalent to township in the Mid-Atlantic and some other regions of the country. "Village" refers to a concentrated settlement within the town.

4 Walker-Kluesing Design Group, *Preservation Guidelines for Municipally Owned Historic Burial Grounds and Cemeteries* (Boston: Massachusetts Department of Environmental Management, 2000).

5 "Ten Most Endangered Historic Resources of Massachusetts," (Boston: Preservation Massachusetts), 2004.

6 "An Introduction to Historic Landscape Preservation," *Terra Firma* 1 (2005); "Caring for Mature Trees in Historic Landscapes, *Terra Firma* 2 (2006); "Identifying and Protecting Historic Roads," *Terra Firma* 3 (2006); "New Models of Stewardship," *Terra Firma* 4 (2007); "Stones that Speak: Forgotten Features of the Landscape," *Terra Firma* 5 (2007); "Commonwealth: The Past and Future of Town Commons," *Terra Firma* 6 (2008); "Taking Action: A Toolkit for Protecting Community Character," *Terra Firma* 7 (2008); "Rooted in History: Preserving Historic Farms," *Terra Firma* 8 (2009).

7 See *http://www.communitypreservation.org*, accessed 6 September 2008.

8 See *http://www.sustainable.org/information/aboutsuscom.html*, accessed 6 September 2008.

Tree Preservation and the Cultural Effects of Climate Change

Rebecca Crew

As climate change occurs, one area historic preservation must address is the changing vegetation in our cultural landscapes. Trees are a dominant feature of many historic landscapes, and as global climate change intensifies, tree habitats will change on a wholesale level. Because trees sequester carbon, increased tree plantings will be part of any viable plan to reduce global warming. Such efforts will inevitably affect historic landscapes, just as inaction would also affect them. Historic preservation professionals will have the opportunities to address these undertakings if they fall under government compliance regulations, and thus we need to assess our current processes of documenting, evaluating, protection, preserving, and restoring historic landscapes to determine if they appropriately respond to the needs of this kind of climate change, while also establishing criteria for sustaining balance between the needs of natural resource preservation and preserving historic landscapes. The historic preservation movement can further its connection to the green movement by considering the cultural aspects of trees—the emotional associations, irreplaceable heritage, and economic realities—and engaging in rhetoric that will humanize the realities of global climate change.

The challenges are numerous; some of them are basic. In many, perhaps most, cultural resources Section 106 compliance processes, architectural historians determine the significance and integrity of a site, and therefore its eligibility for listing in the National Register of Historic Places. Setting is one of the seven aspects of integrity that a resource should retain in order to be considered eligible, and the trees and landscaping that surround a historic building or site are integral to defining the character of the setting. However, state survey forms vary to a considerable degree in their provisions for the documentation of setting. To identify the best practices of historic landscape documentation, I reviewed survey forms from thirty-four state historic preservation offices. I found that that less than 50 percent of this sampling contained sections specifically dedicated to recording the setting or landscape features, although many did list landscape as a thematic classification of significance or allowed description of the setting within the general discussion of the resource. On the other hand, Delaware, Massachusetts, and Ohio offer an entire form for documenting landscapes, and New Jersey uses a separate attachment for the same purpose. California's form offers a flexibility that allows for landscapes, and Alaska requests that the surveyor document the ecosystem of the resource. Michigan includes a chapter on surveying landscapes and farms in its survey manual. However, the great majority of states have no stipulation in their forms for any information about landscape features. Especially with the copious detail that is now generally given to documenting the fabric and design of a building, the setting demands more consideration. The plantings, or lack thereof, are important to understanding the material culture of the resource, perhaps just as much as the kind of fenestration on a building. The National Park Service's *Preservation Brief 36*—"Protecting Cultural Landscapes"—provides guidance on developing plant inventories, noting that cultural landscapes may be as small as a suburban lawn or as large as a 700-acre battlefield.[1] With improved use of GIS, existing plant inventories conducted by natural resource specialists can be incorporated into historic site documentation and historians can learn to interpret what it means if a lawn is manicured with perfectly cut grass or if a house is shaded by evergreens. Recording the size, or age, and type of species at a survey location will lead to a greater understanding of the material culture of our landscapes as well as to recording a baseline for vegetative growing zones.

The scientific community, the general public, the Supreme Court, and now several governmental agencies agree that global warming is a reality. Researchers from the Natural Resources Defense Council and Tufts University produced a chart of American cities, calculating the temperature change they will experience by 2100 if nothing is done to slow down our rate of greenhouse gas emissions. By this estimation, New York will feel like Memphis, Memphis will feel like Miami, and Miami will feel like no comparable city.[2] Along with the environmental devastation that climate change will bring, the impacts to culture and heritage will also be significant. The outcome will extend far beyond routine patterns—not being able to ski, for example, or not bringing out wool coats for winter. It will entail a reconfiguration of where plants and trees will be able to grow, resulting in effects to whole agricultural and forest industries, as well as to the yards of millions of gardening homeowners who may have noticed that the Arbor Day Foundation has already adjusted the United States Department of Agriculture hardiness zone map to reflect current conditions (fig. 4.1). Preservationists should be cognizant of these changes and work toward documenting the landscape heritage that may disappear within our lifetimes.

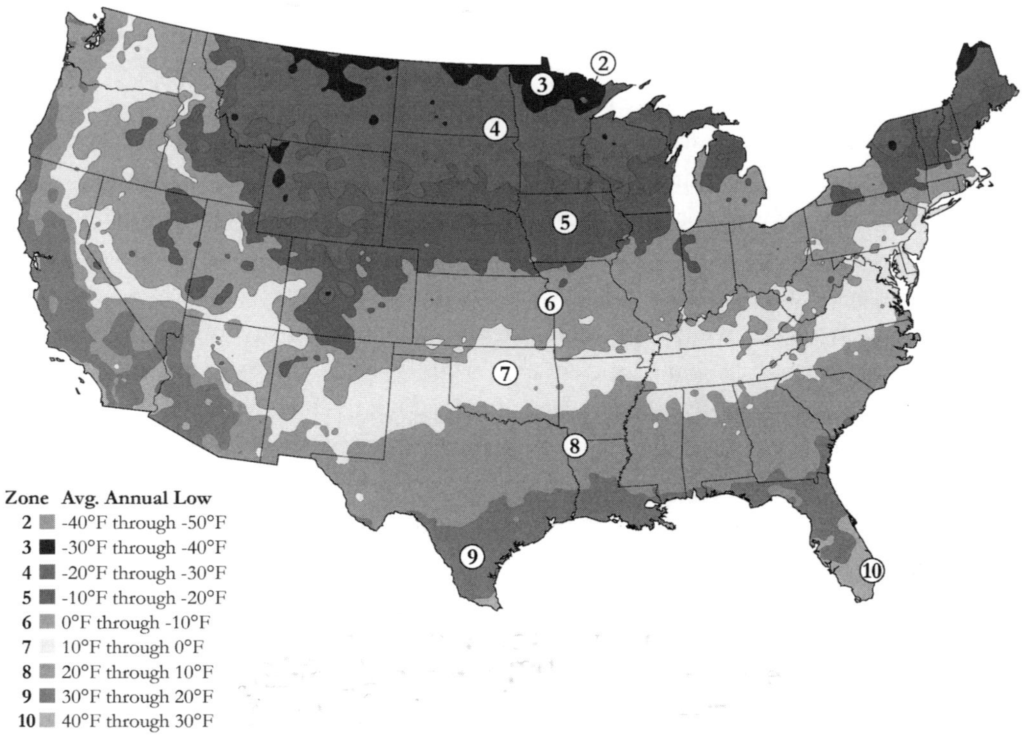

FIGURE 4.1 Hardiness zones map, 2006, National Arbor Day Foundation, showing effects of warming climate on areas appropriate for planting of tree species. Courtesy National Arbor Day Foundation.

The U.S. Forest Service's "Climate Change Tree Atlas" shows the expected change in suitable growing range and likely areas of frequency for 134 tree species of the eastern United States. The maps reveal that some species will have an increased range of suitable climate, while others drift northward or inland, and other species with specific climate condition needs, such as the big leaf magnolia, will have even smaller areas of suitable habitats based on climate conditions. Some interesting findings are that flowering dogwoods, now associated with the upper South, could be most concentrated in New York State, but may also be found in Minnesota as well as southern Florida. White spruce, now found in our northern states, could be suitably grown in Arkansas. Bald cypress, a tree of the Deep South, might be grown up the East Coast as far north as Maine. Live oaks, now found mostly in southern coastal plains, could expand well into the piedmonts of South Carolina and Georgia, and into the coastal plain of states like North Carolina. Paw paws, now common in the Ohio River Valley, would be suitable in New England; and the pecan, the state tree of Texas, shows a huge potential area of growth, expanding to most of the eastern states.[3] More than half of all states may no longer offer suitable habitats for their official state trees and/or flowers.[4]

We have already seen the beginnings of climate change through the increased frequency of dramatic weather events. In 2005, hurricanes Katrina and Rita, two storms of the kind that will be more common as global warming increases, destroyed approximately 320 million trees in two states.[5] Mostly yellow pines, these trees grow quickly compared to slow-growing native pines. By being so fast-growing, they are not as strong as slower-growing trees, and they have shorter life spans. Ancient cypress trees that are native to the same region are hundreds of years old; they remained relatively unharmed, while the fallen pines were left to decompose, contributing greenhouse gases to the atmosphere. Despite the relative youth of the yellow-pine trees, the loss of them in such magnitude altered the landscape of entire communities. The selection of replacement trees should consider new weather patterns but also respect the natural heritage of the region.

Besides the heavy rain events associated with global warming, more extreme temperature changes are another risk. In fruit-growing regions of the southern United States, warmer early springs will bring on earlier fruit blossoms, which will then be at greater risk for damage from late freezes, causing severely reduced fruit harvests. The increased risks of weather to affect fruit production may further reduce the number of individuals that choose to farm or to continue to farm.

Similar changes are seen in Vermont, New Hampshire, and other New England states, where maple syrup is a heritage product. Maples have to be cold enough to produce maple syrup, and in recent years the maple syrup yields have been reduced. Syrup producers believe at this rate they will no longer be able to tap enough maple syrup to continue economic viability.[6] Eventually, the sap supply may decrease so that no maple syrup at all will be produced in the United States, and it will all be imported from Canada.

While importing maple syrup from Canada is mostly an economic and emotional issue, international trade is a factor that poses risks to trees through the spread of invasive plants and pests. The introduction of new species to new countries is generally avoided because non-native plants can become invasive. While some exotic species thrive without taking over, many others require unreasonable amounts of resources, especially water, to keep happy during droughts. Boxwoods, forsythia, honeysuckle, wisteria, and many of the plants found in historic gardens of the South are not native and are hard to consider exotic, but the non-native species of honeysuckle and wisteria are actually considered invasive. According to the National Wildlife Foundation, climate change will encourage the spread of invasive species, particularly kudzu, garlic mustard, purple loosestrife, and Japanese honeysuckle.[7]

In 2002, the emerald ash borer, an exotic Asian beetle, appeared in Michigan. Since then the invasive species has destroyed tens of millions of ash trees, resulting in U. S. Department of Agriculture quarantines and causing economic loss to municipalities, property owners, nursery operators, and forest-product industries. While not directly a result of global warming, the spread of the emerald ash borer may be accelerated by the process as a result of causing stress to ash trees so that they are less resistant. A warming climate can also quicken the beetle's reproductive cycle. The emerald ash borer is the latest in a series of pests and fungi that have hit the United States during the last century and resulted in dramatic change to the landscape.

In 1904, when the chestnut blight fungus from Asia was identified in New York, the American chestnut was the dominant forest species throughout the eastern United States. Chestnut blight virtually eliminated the tree from the American landscape; few specimens of the species now grow more than twenty feet tall, unable to reach reproductive maturity before being killed by the blight fungus. Similarly, Dutch elm disease, caused by the elm bark beetle from Asia, appeared in the United States in 1928 and wiped out the popular street tree in hundreds of American towns and cities. A significant countermeasure began in 2005 when the U.S. Department of Agriculture and the National Park Service released the Jefferson elm cultivar, developed in the 1930s for elms enframing the Mall in Washington, D.C., to commercial trade, allowing for the return of elms to the American landscape (fig. 4.2). During the 1920s, too, another Asian insect, the woolly adelgid, began attacking hemlocks in the Pacific Northwest. Three decades later, it appeared near Washington, D.C. and Richmond, Virginia, beginning its attack on the eastern hemlock, which vividly continues today in Shenandoah National Park and the southern Appalachians (fig. 4.3).

Meanwhile, the Florida torreya pine, perhaps an early example of global warming's threat, is on the edge of extinction in its tiny historic range in Florida, with perhaps 1,000, mostly immature, living specimens. The leadership of an organization called Torreya Guardians believes the species would thrive in the southern Appalachians, due to the habitats of closely related species and to the health of a stand planted in Biltmore Forest in 1939. Torreya Guardians have conducted controversial "assisted migration" by planting specimens ten times further than it spreads naturally via transport by squirrels. While the Florida torreya is not likely to become invasive in the southern Appalachians, it could essentially take the place of the hemlock, as more and more hemlocks are lost to the woolly adelgid.[8] Because the Florida

FIGURE 4.2 and 4.3 (Left) Jefferson elms lining Mall, Washington, D.C., planted in 1930s. Photograph by Jet Lowe for Historic American Landscape Survey, 2006. Library of Congress, Prints and Photographs Division, P&P, HALS, DC, 7-1. (Right) Cabin at Camp Hoover, Shenandoah National Park, Virginia, showing impact of woolly adelgid on eastern hemlocks. Photograph by William A. Faust for Historic American Engineering Record, 1996. Library of Congress, Prints and Photographs Division, P&P, HAER, VA-119-69.

torreya grew in the southern Appalachians during the last ice age, the debate of assisted migration re-examines the notions of plant preservation, conservation, and restoration, as well as the historic preservation tenet of a period of significance.

Thus, in this era of global climate change, when trees have been identified as carbon neutralizers, initiatives to plant trees in massive numbers are likely in order to diminish the threat of extreme weather that takes the form of sea-level rise, desertification, stronger hurricanes, cyclones, and typhoons, more flooding, more droughts, more ice storms, and more blizzards (fig. 4.4).[9] While this extreme weather threatens built structures and their landscape settings, the wise planting of trees may diminish the threat. Greenhouse gas emissions could be significantly offset by changes in agricultural practices that include the foresting of marginal agricultural land.[10]

Already, environmental and ecological groups, as well as governmental and commercial organizations, have developed programs and methods that will help the future to flourish with trees. Urban forestry programs ensure that when the existing mature trees reach the ends of their lives, younger trees will be ready to replace them.[11] Many municipalities have enacted tree

FIGURE 4.4 Norway spruce trees, Lake Montebello, Baltimore, Maryland, showing impact of recent snowstorms. Photograph by Rebecca Crew, 2010.

FIGURE 4.5 Recently constructed infill row houses in Canton neighborhood, Baltimore, showing young street trees planted in modestly sized pits. Photograph by Rebecca Crew, 2010.

preservation ordinances in which trees with a certain diameter must be replaced when cut down for development. City arborists have developed policies on selecting species with long life spans. As an example, in 2005, Baltimore set a tree canopy goal to increase tree coverage from twenty to forty percent citywide, as this increases ecological health as well as property values.[12] The Baltimore Ecosystem Study found, not so surprisingly, that tree coverage is greatest in neighborhoods that match the market segmentation of economically and educationally elite neighborhoods with mostly white inhabitants.[13]

With seventy-one local or National Register-listed historic districts, a major portion of Baltimore City is officially "historic," so tree canopy coverage does not correspond so much to the age of neighborhoods as to their density. Historic areas that were once home to factory workers were constructed to house many people in small spaces; street sidewalks were relatively narrow, with less room for trees, than in neighborhoods established for more affluent Baltimoreans, where even though the row house form might predominate, the historic streetscape included trees.[14] Today, high tree canopy and historic district designation are both factors that contribute to higher real estate values. However, historic neighborhoods with simple, working-class houses that lack ornamentation are less valued as "historic" properties. For example, new homeowners in the Canton neighborhood, who are considerably more affluent than their predecessors, often opt for gut rehabilitation, with exposed brick walls and

FIGURE 4.6. Park Avenue, Bolton Hill neighborhood, Baltimore, showing mature street trees planted in generously sized pits. Photograph by Rebecca Crew, 2010.

modern fixtures. With gentrification has come the planting of street trees (fig. 4.5). In elite neighborhoods of the nineteenth century, such as Mount Vernon and Bolton Hill, cautious restorations are complemented by uniformly placed, mature trees in iron-framed street-tree pits (fig. 4.6). Other city neighborhoods range from a tree canopy as low as .38 percent in Highlandtown (known for its Formstone-clad, working-class row houses) to ones as high as 54 percent in Mount Washington, a streetcar suburb that has its own arboretum (figs. 4.7, 4.8). Some Baltimore neighborhoods have hosted anti-tree sentiments for decades, blaming trees for property damage of various kinds.[15] While preservationists could argue that trees can damage historic resources, the more sustainable option is to support urban forestry efforts. Because municipal private monies are the primary funding sources for urban forestry, the resulting projects are generally not subject to Section 106 review, and historic preservation compliance is rarely required. Still, preservationists may have a role guiding appropriate plantings, through measures such as historic district guidelines and advocating for the replacement of trees in historic property management (fig. 4.9).[16]

FIGURES 4.7 and 4.8 (Top) Highlandtown neighborhood, Baltimore, showing typical absence of street trees. Photograph by Rebecca Crew, 2010. (Bottom) House in Mount Washington neighborhood, Baltimore, showing abundance of mature trees and of English ivy. Photograph by Rebecca Crew, 2010.

FIGURE 4.9 Beverly Hills neighborhood, Baltimore, showing newly planted London plane tree (foreground) replacing sycamore of same vintage as those in background. Photograph by Rebecca Crew, 2010.

Historic preservation is more likely to play a role in tree-planting initiatives with federal funding, such as promising efforts by state transportation agencies. Roads and trees have long been intertwined, as early twentieth-century landscape architects, planners, and politicians argued for the planting of street trees for shade and food, the building of true parkways shaded by trees for health and aesthetic enhancement. Even then, native plants were recommended, and in the mid-twentieth century, Lady Bird Johnson promoted the idea of beautification of our highways, an initiative that included growing native wildflowers along highways and in medians.[17] The Federal Highway Administration has launched a pilot project for quantifying the amount of carbon that can be sequestered, using native vegetation management on Department of Transportation lands.[18] Environmental planners now use the term "green infrastructure" to make sure that green spaces connect, creating and preserving large habitats for various species. Green infrastructure is coupled with the green highways concept, which attempts to improve the environment around highway projects by measures such as retaining streams' riparian buffers and recycling road construction materials.[19]

Another trend is the road diet, developed as context-sensitive design, which actually narrows the width of an existing roadway, providing space for street trees in sidewalks and medians. Road diets result in traffic calming, bicycle lanes, and wider sidewalks that promote vibrant street-life economy, as well as environmental benefits, such as less runoff, more efficient vehicle use, and cooler street temperatures.[20] Through transportation-related tree plantings, green infrastructure planning, and road diets, historic preservationists have opportunities to contribute cultural considerations to these innovative efforts.

The trends in environmental planning are significant for preservationists. A position paper from the Advisory Council on Historic Preservation recommends that in integrating historic preservation and transportation planning, "consider historic properties as part of a broader environmental ecology," meaning that the firm boundaries required by Section 4(f) compliance may prevent opportunities for overlap between natural and cultural values.[21] Recent case law trends indicate that climate change should be included in environmental review under the National Environmental Protection Act for particular projects. Thus, it is time to consider how climate change affects historic resources, and for a reassessment of current compliance requirements in order to better incorporate cultural landscape preservation. The first step is better documentation of existing landscape conditions, and to correlate existing conditions with what is predicted to be at stake in climate change. Through historic assessments, we will gain a fuller understanding of the significance of these cultural landscapes. While we will not find all cultural landscapes eligible for the National Register, this does not mean they are without preservation value, just as there are environmental and cultural benefits to preserving buildings that are determined not eligible for the National Register. By protecting what we now have, we provide a firmer foundation for the future. Without knowledge of and reverence for the living elements in our landscapes, future generations are unlikely to put forth effort towards the technological innovations and political changes needed to reduce climate change impacts. However, through the combination of historic and natural resource preservation, cultural landscapes may remain sustainable.

Notes

1 Charles Birnbaum, "Protecting Cultural Landscapes: Planning Treatment and Management of Historic Landscapes," *Preservation Briefs* 36 (1994); see *http://www.nps.gov/history/hps/tps/briefs/brief36.htm*.

2 Frank Ackermann and Elizabeth A. Stanton, "The Cost of Climate Change: What We'll Pay if Global Warming Continues Unchecked," Natural Resources Defense Council, May 2008; see *http://www.nrdc.org/globalwarming/cost/cost.pdf*, accessed 25 August 2008.

3 "Climate Change Atlas: A Spatial Database of 134 Tree Species of the Eastern USA," U.S. Department of Agriculture, Forest Service; see *http://www.nrs.fs.fed.us/atlas/tree/tree_atlas.html*, accessed 12 November 2008.

4 "Gardeners Guide to Global Warming," National Wildlife Federation; see *http://www.nwf.org/Get-Outside/Outdoor-Activities/Garden-for-Wildlife/Gardening-Tips/Gardeners-Guide-to-Global-Warming.aspx*, accessed 28 February 2010.

5 Marc Kaufman, "Katrina, Rita Caused Forestry Disaster," *Washington Post*, 16 November 2007, A10.

6 Pam Belluck, "Warm Winters Upset Rhythms of Maple Sugar," *New York Times*, 3 March 2007, A1.

7 Patty Glick, "The Gardener's Guide to Global Warming: Challenges and Solutions," National Wildlife Federation, 2007; see *http://www.nwf.org/gardenersguide/Gardeners_Guide.pdf*, accessed 12 November 2008..

8 Michelle Nijhuis, "Taking Wildness in Hand: Rescuing Species," *Orion* 27 (May-June 2008): 64-78; see *http://www.orionmagazine.org/index.php/articles/article/2966*.

9 Anthony Janetos, interview with Ray Suarez, "Online NewHour," PBS, 28 May 2008; see *http://www.pbs.org/newshour/bb/science/jan-june08/climate_05-28.html*, accessed 14 July 2008.

10 "Agricultural and Environment Initiative," American Farmland Trust, 2008; see *http://action.farmland.org/site/DocServer/AFT-Agriculture-and-Environment-August2008.pdf?docID=1341*, accessed 30 December 2008.

11 Arbor Day Foundation, "Tree City USA," n.d.; see *http://www.arborday.org/programs/treeCityUSA/index.cfm*, accessed 31 January 2008.

12 Myra Brosius, "TreeBaltimore Urban Forest Management Plan," Baltimore City Forestry Division, 2007; see *http://www.ci.baltimore.md.us/government/recnparks/downloads/TreeBaltimore%20Urban%20Forest%20Management%20Plan.pdf, accessed 25 August 2008.*

13 Baltimore Ecosystem Study, "Demographic and Socioeconomic Projects," 2008; see *http://www.beslter.org/frame4-page_3d.html*, accessed 25 August 2008.

14 Mary Ellen Hayward and Charles Belfoure, *The Baltimore Rowhouse*. (New York: Princeton Architectural Press, 1999).

15 Brosius, "TreeBaltimore."

16 John Milner Associates, "Draft Historic District Guidelines Update," Baltimore City Commission on Historic and Architectural Preservation, January 2008; see *http://www.ci.baltimore.md.us/government/historic/downloads/DesignGuidelines/Chapter%206.1%20%201-8-08.pdf*, accessed 25 August 2008; City of Gaithersburg Historic District Commission, "Brookes, Russell, and Walker Historic District Design Guidelines," n.d.; see *http://www.gaithersburgmd.gov/documents/design_guidlines.pdf*, accessed 25 August 2008; Patricia Patula, ed., "Chestnut/Meem Historic District Guidelines," City of Gaithersburg, Historic District Commission, December 1998; see *http://www.gaithersburgmd.gov/documents/chestnut_meem.pdf*, accessed 25 August 2008; "Green Home Tips," National Trust for Historic Preservation, January-February 2008; see *http://www.preservationnation.org/issues/sustainability/green-home-tips.html*, accessed 30 December 2008.

17 "Our Environmental First Lady," Lady Bird Johnson Wildflower Center, University of Texas, Austin, n.d.; see *http://www.wildflower.org/environmental_first_lady/*, accessed 25 August 2008.

18 Federal Highway Administration, "Carbon Sequestration Along Highway Rights of Way: Piloting a Concept," Successes in Stewardship newsletter, September 2008; see *http://www.environment.fhwa.dot.gov/strmlng/es4newsltrs.asp*, accessed 28 February 2010.

19 "Green Highways Partnership: Stewardship, Safety & Sustainability," Global Environment & Technology Foundation, 2007; see *http://www.greenhighways.org/*, accessed 25 August 2008.

20 Jennifer Rosales, "The Road Diet Handbook: Setting Trends for Livable Streets," William Barclay Parsons Fellowship Monograph 20, Parsons Brinckerhoff, New York, 2006.

21 Carol Legard and Don Klima, "Integrating Historic Preservation and Transportation Planning: A Five Point Plan for Making It Work," Advisory Council on Historic Preservation, draft, 27 September 2008. Section 4(f) of the U.S. Department of Transportation Act of 1966 protects publicly owned public parks, recreation areas, and wildlife and waterfowl refuges as well as historic sites of local, state, or national significance, from conversion to transportation uses.

Awbury Arboretum: A Quaker Legacy Sustaining an Urban Community

Valencia Libby

The mission of Awbury Arboretum is to preserve and interpret Awbury's historic house and landscape, thereby connecting an urban community with nature and history (adopted 2001).

During 2006 and 2007 Philadelphia was often in the national news due to its homicide rate, one of the highest of any American city, five times the national average. Unchecked government corruption and economic backsliding since the 1970s had left the "City of Brotherly Love" a city in crisis. In 2008 four Philadelphia police officers were killed. Compared to problems of this social magnitude, the preservation of historic landscapes and buildings seems irrelevant, and perhaps they are unless they are being utilized to serve current populations. Set against this bleak background, Awbury Arboretum has been an inspiring experiment in social and environmental sustainability since 1916.

Awbury Arboretum is a fifty-five-acre oasis in the northwestern section of Philadelphia known as Germantown. Its picturesque landscape was developed from the mid-nineteenth to the early twentieth centuries by several generations of the Haines and Cope families, affluent members of the Quaker community.

The arboretum also forms the matrix of Awbury Historic District (recognized in 2001), which contains thirty-one historic buildings, mostly houses arranged in irregular clusters (fig. 5.1).[1] The combination of public arboretum, historic landscape, and historic houses, some of which remain occupied by their owners, makes for a distinctive ensemble, but one that is very complicated to manage. The location within a densely developed, low-income neighborhood renders the task even more challenging. However, the Awbury Arboretum Association has imaginatively responded to these conditions since 1982.

Awbury began as a pastoral retreat for the Haines and Cope families of Philadelphia. In 1849 John S. Haines (1820–86) purchased the first parcel to build a summer estate at the higher elevations beyond the city. The land was gently rolling, open fields with few trees or structures on it. It commanded a view to the east over the surrounding area. In 1852 his father-in-law, Henry Cope (1793–1865), joined him, buying twenty-three adjacent acres and naming the estate, "Awbury," in memory of the family's ancestral home near Avebury,

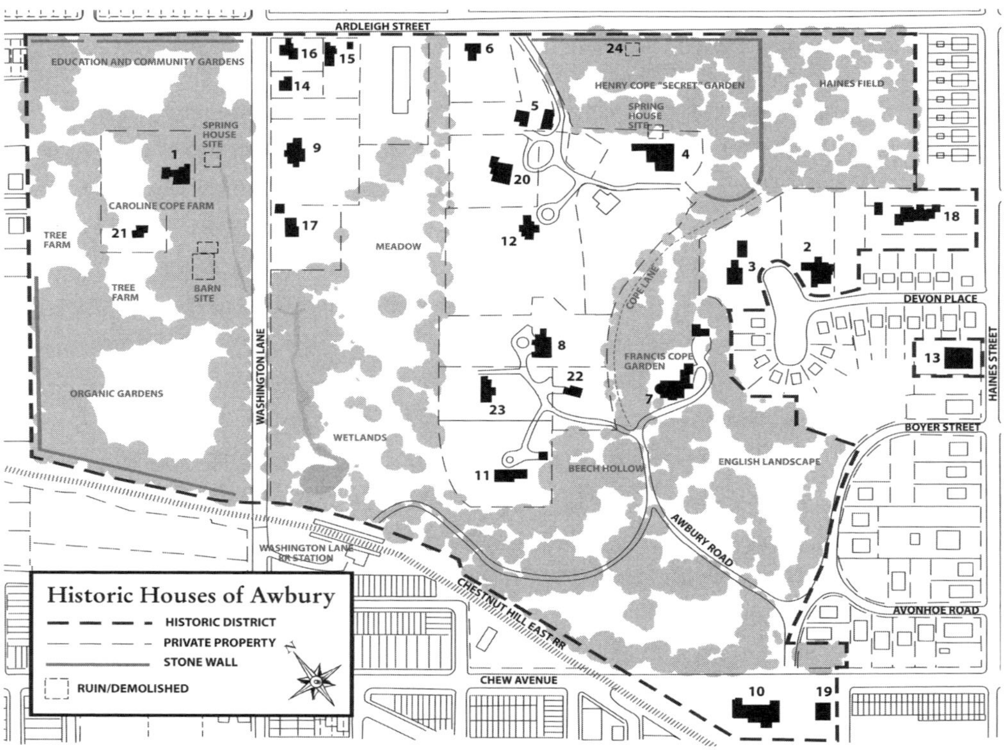

FIGURE 5.1 Awbury historic district, site plan, showing John and Mary Cope Haines house (2) and carriage house (3) (both 1849–50); Henry and Rachael Cope house (4), carriage house and stable (5), and tenant house (6) (all 1852–53); Francis and Anna Cope house (7) (1860–62); and Caroline Cope house (1) (1793, purchased 1885, additions 1921). From Gay G. Johnson and Mark R. Sellers, *The Historic. Houses of Awbury*, 2008.

FIGURE 5.2 Henry Cope house. Photograph by Reuben Haines III, ca. 1900. Courtesy Awbury Arboretum Archives.

England. Henry Cope built a large summer house and a double tenant house for his extensive family and continued to buy up parcels of land until he owned approximately forty-two acres. In 1860–62 he also built in a handsome stone house centrally located near the top of a hill for his oldest son, Francis R. Cope (fig. 5.2). In 1854 Germantown was consolidated into the City of Philadelphia and by 1863 a new branch line of the Philadelphia & Reading Railroad located a station near Awbury's southern perimeter, making it convenient to commute into the city. Awbury's residents began to convert their summer houses into year-round homes. Succeeding generations acquired land to group their houses around those of their relatives within the evolving landscape of woodlands and fields that were enjoyed by all.[2]

Few details are known about the earliest period of landscape development from 1850 to 1870; no primary documents have been found, except for the receipt book of William Saunders, a professional landscape gardener from Scotland, who recorded working for Awbury between 1856 and 1859. The Germantown historian, Edwin C. Jellett, attributed the early landscape design of Awbury to Thomas P. Cope (1823–1900) who worked extensively on the planting of the estate prior to Saunders's arrival.[3] Other family memoirs support this claim. Elizabeth S. Cope (1848–1937) described the estate during her childhood as having a carriage drive and paths "but there were no large trees except for a few old gnarled apple trees and two old black cherries ... Beyond stretched fields and open country for ten miles where we could trace the blue hills on the far side of the Delaware River."[4] Historian Veronica Aplenc noted in Awbury's cultural landscape report that there are no primary documents defining boundaries, drives, or entrances for the first twenty years of the estate's history.[5] Furthermore, one of the major problems in writing an accurate history of Awbury is the lack of a complete title search tracing numerous additions and subtractions to the estate throughout its history, even after the establishment of the arboretum.

By the late nineteenth century, the three major landowners at Awbury were Henry Cope's two sons, Francis R. and Thomas P. Cope, and John S. Haines. Their contiguous lands extended between four major public roads: Chew Avenue to the south, Sullivan (now Ardleigh) Street to the north, Washington Lane to the west, and East Walnut Lane to the east. Caroline E. Cope bought an eighteenth-century farmhouse with land on the northwest side of Washington Lane in 1885 to be near her family. The owners continued to develop the landscape character of the estate with the careful placement of groves of trees, a curvilinear drive system and "The Boardwalk," a plank-lined path between silver maples that led from Francis R. Cope's house downhill to Chew Avenue. The area outside of Awbury was changing very slowly. McNabbtown, a shantytown for immigrants built in 1870 on a small triangle of land adjacent to the railroad line, abutted Awbury's southern perimeter. The first clusters of row houses and commercial enterprises were built along Chew Avenue near Washington Lane and the railroad station.

By the early twentieth century, Awbury's landscape had matured with a distinctly picturesque character, elements of which are still visible today. Period photographs show the intentionally naturalistic composition, extensive lawns and meadows, open vistas framed by groves of trees and shrubs, a narrow drive curving through woodland, and informal footpaths connecting dispersed homes surrounded by lawns (fig. 5.3). Individual owners made ornamental and kitchen gardens near their homes and built large vine-covered arbors for shade (fig. 5.4). Awbury became a serene enclave from whose long verandas one could glimpse sailing ships on the Delaware River. In *Gardens and Gardeners of Germantown* (1914), Jellett praised Awbury: "It is like a great park, abounding with walks, drives, rare shrubs, trees, and richly stored gardens ... of all the experiments in so-named community life, to me, 'Awbury' is the most practical and beautiful."[6] The generation of children raised at Awbury during this period were often were reminded, "Enjoy this beauty while you have it; for it cannot last long."[7]

Founding an Arboretum

During the 1910s and 1920s, the area surrounding Awbury developed rapidly as immigrants and others flooded into the city. Farms and country houses for the affluent gave way to lines of row houses on small rectangular plots for working families. Commercial enterprises encroached on the pastoral scene. To the south of Awbury, closer to the city, the density of development was greatest. To the north, semi-detached and detached houses were built on larger

FIGURES 5.3 and 5.4 (Top) Awbury, view of grounds looking southeast from highest elevation, showing urban development in background. Photographs ca. 1910s. Courtesy Awbury Arboretum Archives. (Bottom) Henry Cope house garden, showing wisteria-covered arbor. Photograph by Reuben Haines III, ca. 1900. Couresty Awbury Arboretum Archives.

parcels. Land values increased and each edition of the *Philadelphia Atlas* showed extensions of public roads dividing up Awbury's acreage, assuming that it would become available for development in the future. Seeing this change, Caroline E. Cope (1840–1944) and her nephew-in-law, William Draper Lewis (1867–1949), a prominent attorney and dean of the law school at the University of Pennsylvania, began a discussion early in 1916 about how to preserve Awbury's pastoral character. Lewis may have been the driving force behind these negotiations. Caroline E. Cope agreed to donate funds for the purchase of almost twenty acres of Awbury to the City Parks Association (CPA), a citizens' organization founded in 1888 to establish neighborhood parks throughout the city. Four of Cope's female relatives approved of the plan, raising an endowment of almost $78,000. Their legal agreement with the CPA, signed on 27 November 1916, specified that the land and funds placed in trust would be used for the purpose of maintaining "a public park and arboretum and furnishing educational facilities in connection therewith."[8] No mention was made of historic preservation in the original documents. Over time other family members lent their support with additional contributions.

The Copes' foresight to incorporate the land surrounding their residences as a public arboretum did protect it from urban development. The Copes believed that they were conserving a way of life in close touch with nature by setting aside the grounds of Awbury and by sharing it with the public. This viewpoint is in keeping with the importance placed upon nature and the study of natural sciences within Quaker culture. While many of Philadelphia's historic houses and parks, including the Morris Arboretum, were given to the public by Quaker owners, the Copes continued to own and live in their dwellings. When they no longer could reside there, they could sell the buildings. Thus the pattern of private ownership in Awbury evolved but it did not end. "They were very canny," said Awbury resident Spencer E. Coxe (b. 1917), as it reduced the tax burden on the family as well as protecting the quality of life they had enjoyed for many decades in the midst of a growing neighborhood.[9] Although not always easy to come to terms with other residents, in general, people have seen the advantage of living within a large, landscaped park in the city.

The CPA was to maintain the arboretum using the endowment income, directed by a seven-member committee, most of whom were family members or other residents. In the original agreement, the family also gave funds to purchase McNabbtown, an adjacent shantytown, and have it demolished, thus removing one eyesore. William Draper Lewis had already begun a correspondence with the landscape architect Arthur Wescott Cowell regarding the design of a botanic garden with specialized collections. In 1918 Cowell supplied the committee with a "general plan" for the arboretum's physical development, which he revised in 1919. He demonstrated sensitivity to the predominant landscape character of the estate by incorporating groups of compatible trees and shrubs around the open spaces in the park. He extended the existing circulation system and showed the former McNabbtown area converted into a formal rose garden, distinct from the otherwise naturalistic areas.[10] Cowell revealed his understanding of the arboretum landscape in 1918 as "a restful and charmingly rural scene made up of pleasant glades and plantations ...."[11]

Cowell continued to revise and provide planting plans for Lewis and the committee through the 1920s, but most of his plans were never realized. Some family members and residents resisted the idea of making dramatic changes to the landscape. They were much opposed to the removal of large trees, even to make space for more desirable botanical species. The one landscape feature from Cowell's general plan to be implemented was a series of three ponds fed by a natural spring in a low valley along the east side of Washington Lane.[12] Constructed in 1919, this naturalistic ensemble became a popular feature of Awbury. The arboretum committee also followed Cowell's suggestions for improvements to the perimeter of the property and for increasing the variety of ornamental tree and shrub species. Otherwise, the overall landscape remained very similar to its estate appearance in 1916.

Growing Problems

The arboretum evolved slowly under the CPA and the committee. The public began to visit the site and maintenance was carried out during the 1920s. In 1923 the heirs of Francis R. Cope donated the family home to the arboretum. The educational mission of the arboretum was left to individual motivation as no formal programs were provided. In 1933 they hired their first director, possibly because of new demands on the arboretum. The open space at Awbury became a significant asset to the neighborhood for food gardens during the Depression. Although the committee gently objected to this use as it was not part of the arboretum's mission, they granted use of "Lilac Hill" to the Germantown Community Center for garden space in January 1933.[13] Three years later they expanded the area available for gardens for the unemployed on ground near the railroad.

This first organized community use of the property initiated what became a long tradition of community gardening at Awbury. WPA crews were employed in 1933–34 to build sections of stone wall along Awbury's perimeter on Washington Lane and Ardleigh Street that unintentionally added to an outer aspect of inviolability. The neighbors began to call Awbury the "Quaker Kremlin."[14] Vandalism, littering, vagrancy, and petty crime were chronic problems in the neighborhood's open spaces. In 1938 Philadelphia's police superintendent warned the Copes that the park lay "in the center of a very rough neighborhood composed of Italians and Negroes. It is bound on one side by a very large tract of vacant property that is full of hobo jungles."[15] The tract he referred to had been owned by the Cope-Nolan family, deeded to the CPA, and then sold to the city in 1936. A portion of it eventually became a city park.

Problems only grew worse after World War II, with a rising crime rate, spreading economic blight, and the white middle-class flight to new suburbs. While the neighborhood to the north of Awbury changed from a white to a black middle-class community, the area to the south and west became increasingly impoverished and depressed. Simultaneously, the CPA failed to maintain the arboretum. Despite the advisory committee's efforts, it lacked direction and there arose questions of financial mismanagement. Family members dwindled and houses were sold to new residents who found themselves picking up trash, cutting down weeds, policing the grounds, and chasing out prostitutes.[16] Given its urban context, however, the crime was not as threatening to residents as in other parts of the city. In 1977 it appeared that Awbury had reached a financial crisis, with the CPA threatening to balance the budget by selling off land.[17] Strong opposition to this move came from Awbury's residents, and as a result, the advisory committee was forced to look very closely at its financial prospects for the future.

A development committee led by Henry Scattergood in 1978 proposed three scenarios for the arboretum's future with their estimated costs. It could become a "neatly maintained" park, the least expensive option. It could become a landscape-training facility for neighborhood youth, the most expensive option. Or it could be the neighborhood environmental center for youth and adult education. The latter two alternatives would require additional grant funding, which was not a dependable source of income, as Scattergood warned. Only raising the endowment would secure Awbury's future and its programs. He also proposed several other income-generating ideas such as a city campground for traveling campers.[18]

All three key alternatives would be possible in time, but the CPA again failed to take action. The final straw was the neglect of the Francis R. Cope house donated to the arboretum in 1923. In 1980 it contained a few rental apartments and an administrative office. The building's distinctive porch, an important architectural feature, was falling apart and the CPA wanted to tear it down. Residents were outraged. If the CPA would do that, then what would they do next? Tear down the whole house? Residents saw the Francis R. Cope house as the embodiment of the arboretum's history.[19] In 1982, a group of residents and concerned neighbors took it upon themselves to find a solution.

Spencer E. Coxe, Awbury resident since 1952, became the group's leader. Coxe had been a conscientious objector during World War II; he had worked for the American Friends Service League in China after the war and had moved to Philadelphia with his wife Elize DeMiranda in the 1950. He became the first executive director of Philadelphia's ACLU, a position he held from 1953 to 1979. Coxe and his wife said that during the time they had lived at Awbury few members of the public used the arboretum except children on their way to school or for sledding in winter. Most of the neighbors thought Awbury was private property—the outer perimeter was dark, heavily vegetated, and "mysterious."[20] The problems facing the arboretum—property vandalism, abandoned stolen cars, illegal dumping, disputed property lines, liability issues, and the deterioration of the Francis R. Cope house—were due to the CPA's incompetence. It was a neglected property. Coxe's group proceeded with careful investigation and thoughtful discussion characteristic of the Quaker process. In 1984 the new group of stakeholders incorporated as the Awbury Arboretum Association and took control of the arboretum's management. They redefined the direction Awbury would take over the next twenty-five years to become a vital community resource.

The association first tackled long-ignored issues of landscape and building maintenance, fund-raising, capital improvements, staffing, and restoration of the Francis R. Cope house, with a phased, five-year plan. They kept the original vision of Awbury's Quaker founders central to their plan—an arboretum, with expanded services to the community. Awbury would not exclude the neighbors and become an insular retreat.[21] Understanding community needs was a priority. Once the association membership had decided on its goals, they acted to fulfill them. Within the first year, when the William Penn Foundation turned down their grant application to restore the Francis R. Cope house, they raised half the necessary amount of $130,000 themselves. They gradually transformed the rental apartments into offices and classrooms for an administrative and educational center. They were the first stakeholders to identify "historical registration" as a priority in 1985.[22] With grant funding, they hired a historian to write the first nomination to the National Register of Historic Places in 1986. It was revised at least twice before they

FIGURES 5.5 and 5.6 (Left) Children at play in Awbury community garden. Courtesy Awbury Arboretum. (Right) Environmental education class at Awbury Arboretum. Courtesy Awbury Arboretum.

received designation as Awbury Historic District in 2001. In 1998, another grant provided funds for a cultural landscape report with recommendations for a master plan.[23] Follow–up action has been limited, however. Today, Awbury's residents seek further protection from ill-considered change through designation as a historic district under the Philadelphia Historical Commission. The application, filed in 2006, is pending review in 2010.[24]

At first the Awbury Arboretum Association tried to work with the CPA, but they found it very difficult. As the executive director, James Alexander, complained to a board member in 1991: "Unless action is taken soon by the CPA, Awbury will shortly revert to its old face of ratty incompetence."[25] Serious liability issues ignored by the CPA caused the association to take over management of the arboretum in the 1990s. The CPA became a vestigial organ that many recognize should now be dissolved.[26]

Community Engagement

Coxe wrote a document in 1987 explaining why the Awbury Arboretum Association had been created and its purpose in which he stated that "one of the first steps was to carry out a survey with the help of faculty and students from Temple University's School of Social Administration to determine the wishes and perceptions of the surrounding community, a densely populated and predominantly black neighborhood."[27] From this and a parallel survey of expert opinion, the association members learned that the community wanted environmental education programs for their children and that they were confused about the purpose of the arboretum, whether it was open to the public or just for the "white gentry" who lived there.[28] Association members wanted the arboretum to serve its urban constituents better. As a result, several programs began to be developed during the mid-1980s.

Gay G. Johnson, one of Awbury's most dedicated residents, initiated a nature camp program for local schoolchildren in the summer of 1984. Her grant request for $8,000 (almost 10 percent of Awbury's annual budget at that time) showed that there were approximately 5,000 children attending four large public schools and fifteen summer day care centers within walking distance of Awbury. Of these children, 99 percent were black, up to 40 percent were from low-income families, and 70 percent tested below the national average. Awbury's program would introduce them to nature in imaginative ways, increase their understanding and enjoyment of the arboretum, and strive to enhance their sense of self worth (figs. 5.5, 5.6). The 1984 nature camp was a resounding success, with 400 children attending over a nine-week period.[29] From then on, year-round environmental education for the young became the mainstay of Awbury's programming. By 2000, the annual number of schoolchildren taking classes at Awbury had increased to 4,000, while the number of families living below the poverty level grew to 60 percent of the population. Today the environmental education program serves an average of 5,000 children per year with the summer nature camp, after-school programs, preschool classes, youth groups, and teacher training.[30]

Community Gardens

Awbury Arboretum has one of the longest histories of community gardening in Philadelphia. During the Depression, the arboretum committee granted use of "Lilac Hill" to the Germantown Community Center to make food gardens in 1933. Committee members expanded the area available for gardens for the unemployed on ground near the railroad in 1936. They wanted to terminate this use at the year's end, but the community's need must have been too great, as the practice continued. During the 1940s, the Germantown Boys Club and two other organizations requested land for vegetable gardens. While some residents felt that the property was an arboretum and this use was in conflict, the onset of World War II saw Victory gardens planted on the northwest tract, land once owned by Caroline E. Cope and deeded to the CPA in 1923. Food gardening continued on the arboretum grounds well into the 1950s.[31] In the early 1970s Penn State's Cooperative Extension helped to establish a large community garden on the northwest tract where it remains, with garden plots for approximately fifty families. The gardens are managed by their own organization with help from Awbury's staff.[32] It is a vitally important link with and source of pride and pleasure for the surrounding community.

The northwest tract of Awbury, once farmland owned by Caroline E. Cope, is separated from the central area of the arboretum by Washington Lane, a major urban thoroughfare, and thus has always been removed from the arboretum's activity. For that reason, the former director, Gerald Kaufman, sought alternative uses for the site that would also improve it. In 2002 Weavers Way Cooperative, a local food cooperative, leased one and one-half acres to start a small organic farm. The organization expanded in 2008, and is building greenhouses and a children's garden. There is also a nursery for city street trees managed by another non-profit group and discussion of educational projects, so that every part of the arboretum will be put to good use. These functions will continue to evolve.

Horticultural Job Training Program

Awbury's most challenging experiment in education began in 1996 when the arboretum's director, Thomas Mishler, learned that funding was available from Philadelphia's Office of Emergency Shelter and Services to assist homeless people. Previously, several job-training programs had been offered to local youths. Mishler created a job-training program that enrolled twelve men who were living in city shelters to work on the grounds of Awbury with the landscape crew. They were taught horticultural and job-related skills intended to lead to gainful employment and independence from the shelter (fig. 5.7). Unemployed women, often welfare mothers, also joined the crew. For some, the exposure to nature was transformative, while for others it was frightening and the daily routine too demanding. Overall, the personal histories of many trainees made long-term employment unlikely; however, the program has had its successes and it has persevered with an average of eleven trainees entering each year. Simultaneously, Awbury offered local teenagers a job-training and environmental-education program along with paid summer internships, all of which were run separately from the adult program.

FIGURE 5.7 Horticulture training program at Awbury Arboetum. Courtesy Awbury Arboretum.

Mishler also established Awbury Landscape Services in 1998 as a for-profit division to generate income from private and public contracts and to employ graduates of the training program. Beginning in 2002, funding from the city became less secure, and the arboretum's succeeding director, Gerald Kaufman, decided to gear the job-training program solely for local high school graduates. He sought and received funding from private foundations for a "youth-at-risk" program. The number of trainees has varied from twenty in 2005 to forty-four in 2006, but many drop out in the first few weeks. The young people work on the grounds for one year, receive classroom as well as field instruction, and travel to see the range of employment available in the green industry. While the cost per graduate is exceptionally high, it remains a daring experiment in social improvement. Awbury Landscape Services is an unquestionable success and an important source of outreach and revenue for the arboretum. The program will continue in 2010 with new partnerships made with other organizations.

Environmental Stewardship

Since the 1980s, Awbury has sought to reclaim the beauty and features of its historic landscape that lay buried in weeds and debris. The board and the administration found no conflict between landscape reclamation efforts that were necessary for the arboretum to function and ecological restoration of diverse natural habitats contained within the fifty-five acres. Philadelphia's long history of environmental restoration carried the arboretum forward with impetus from such renowned landscape architects as Carol Franklin, Leslie Sauer, and John F. Collins. Awbury afforded Germantown a unique opportunity to restore an urban ecology. It made good sense and it made good public relations.

While the cleanup of the woodlands, planted shrubberies, lawns, and fields went forward, Awbury's land manager identified an area requiring special expertise for restoration—the watercourse and wetland. This system consists of two natural springs that drain into channels and flow into man-made ponds and a sunken meadow. Over the years, urban development has altered the water table and increased the amount of stormwater runoff. Awbury's springs have almost dried up and the ponds have silted up. Additionally, heavy rainstorms in recent years sheet across Washington Lane, causing a dangerous situation and flooding the neighborhood's outdated sewer system. This occurrence is repeated many times throughout the city and has led the Philadelphia Water Department to create an Office of Watersheds to serve as the proactive leader in urban watershed restoration and stewardship.

In 2002 Awbury received one of Pennsylvania's "Growing Greener" program grants for $42,000, made available through the state's Department

FIGURE 5.8 Awbury Arboretum staff member planting native trees as part of the 2003 watershed restoration project. Courtesy Awbury Arboretum.

of Conservation and Natural Resources. The funds were used by Awbury, in partnership with the Philadelphia Water Department, to develop Phase 1 of a master plan for a stormwater management system. This phase included "removing invasive plants along waterways and replacing them with native plants, enhancing the meadow, daylighting a stream, and diverting stormwater into an infiltration swale (or bioswale) along Washington Lane." Construction began in March 2003 and was completed with additional grants of $25,000 from the state and $5,000 from the Garden Club of America (fig. 5.8). Further restoration of Awbury's pond and stream system still needs to be carried out with a view to the historic cultural landscape.[33]

The need for public education about watershed protection is a very important aspect of this challenging project, as it will be utilized as an environmental model by both the Philadelphia Water Department and the Tookany Watershed Association. Throughout the city, the Office of Watersheds has helped to establish neighborhood organizations to safeguard numerous water systems feeding into its major rivers. Kaufman invited the Tookany Watershed Association, a young non-profit, to make its headquarters in the Francis R. Cope house, thus fostering another collaboration. His strategy of building a network of partnerships between the arboretum and other organizations, large and small, private and public, is a promising way to secure Awbury's future and will be continued.

The Cope and Haines families founded Awbury Arboretum with a clear purpose in mind framed by a well-conceived and straightforward legal structure. Despite various threats to Awbury over the past ninety years, that purpose has remained strong due to succeeding generations of residents and stakeholders committed to the original vision. There is a distinctly Quaker approach at work here where process is as important as product; nature study, historic preservation, social service, and public education are complimentary threads of a single organization. The original family commitment to place personal assets in trust for the public has succeeded and has been of mutual benefit. As Spencer Coxe noted, "The Copes were 'doing well by doing good'." The financial management and maintenance of Awbury has been and will continue to be the greatest challenge; however, Awbury remains an important case study in sustainability and an outstanding asset to its community. ❁

Notes

1 See district map in Gay G. Johnson and Mark R. Sellers, *The Historic Houses of Awbury* (Philadelphia: Awbury Arboretum Association, 2008).

2 Veronica Aplenc, "Cultural Landscape Report, Part 1: Site History, Exiting Conditions and Analysis, Recommendations for the Master Plan and Beyond of the Awbury Arboretum," 1997, Awbury Archives.

3 Edwin C. Jellett, *Gardens and Gardeners of Germantown* (Philadelphia: Site & Relic Society of Germantown, 1914), 329. Saunders was in Philadelphia only for several years—1854 to 1860.

4 Elizabeth S. Cope, *Awbury Memories* (Philadelphia: Awbury Arboretum Association, 1994), 15.

5 Aplenc, "Cultural Landscape Report," 24.

6 Jellett, *Gardens and Gardeners*, 329.

7 Mary C. Scattergood, *A History of Awbury* (Three Oaks, Mich.: Henry Regnery, 1972), 36.

8 Copy of 27 Nov. 1916 legal agreement between Donors and CPA, Awbury Arboretum Binder, Awbury Archives.

9 Interview with Spencer E. Coxe, Philadelphia, 11 November 2008.

10 Arthur W. Cowell, "General Plan and Scheme of Botanical Arrangement," original of revised plan May 1919, Awbury Archives.

11 Copy of letter from Arthur W. Cowell to William Draper Lewis, 4 April 1918, Temple University Urban Archives, Philadelphia, Acc. 407, Box 10.

12 See Aplenc, "Cultural Landscape Report," 49-55, for a detailed discussion of Cowell's efforts and changes to the arboretum.

13 Minutes of the Awbury Arboretum Committee, 6 January 1933, Temple University Urban Archives, Acc. 407, Box 10.

14 Dianne C. Gordon, "Awbury to Get Aid from U. of P.," *Evening Bulletin* (Philadelphia), 18 July 1979, Awbury file, Temple University Library, Urban Archives.

15 Aplenc, "Cultural Landscape Report," 56.

16 Interview with Gay G. Johnson, Philadelphia, 15 February 2008.

17 Pat McKeown, "Arboretum in Germantown Facing Financial Crunch," *Public Ledger* (Philadelphia), 28 September 1977, Awbury file, Temple University Urban Archives.

18 Henry Scattergood, "Three Program Alternatives for Awbury Arboretum," 18 August 1978, Box 7, Awbury Archives.

19 Coxe interview.

20 Interview with Spencer E. and Elize D. Coxe, Philadelphia, 11 November 2008.

21 File "Awbury Committee General Correspondence 1984-86," Box 7, Awbury Archives.

22 "Awbury Arboretum: Outline for a Five Year Plan 1985-1989," file "Awbury Committee General Correspondence 1984-86," Box 7, Awbury Archives.

23 See Aplenc, "Cultural Landscape Report."

24 Interview with Mark R. Sellers, Philadelphia, 11 September 2008.

25 Letter James Alexander to John Lawson, 6 September 1991, file "AAA Correspondence 1988-1991," Box 7, Awbury Archives.

26 Sellers interview.

27 Spencer Coxe, "The Awbury Arboretum Association," March 1987, Box 7, Awbury Archives.

28 Ibid.

29 Interview with Gay G. Johnson, Philadelphia, 5 May 2008; Gay G. Johnson, annual reports for the Education Program 1985 to 2000, Awbury Archives.

30 Interview with Linda Brown, Philadelphia, 26 September 2008.

31 Aplenc, "Cultural Landscape Report,"62.

32 Linda Brown, "Natural Resources Education Garden Dedicated," *The Arbor*, Newsletter of the Awbury Arboretum Association, June 2000, Awbury Archives.

33 Interviews with Gerald Kaufman, Philadelphia, 22 May and 4 June 2008; copy of "Environmental Stewardship and Watershed Protection Grant Application 2001" for Awbury Arboretum, Awbury Archives.

Is Brown the New Green? Balancing Water Conservation with Historic Landscapes in the San Joaquin Valley

Karana Hattersley-Drayton

California and its great Central Valley have a Mediterranean climate consisting of wet winters and dry, rainless summers. John Muir poetically wrote in *My First Summer in the Sierra*: "In … California there are only two seasons—spring and summer. The spring begins with the first rainstorm, which usually falls in November. In a few months the wonderful flowery vegetation is in full bloom, and by the end of May it is dead and dry and crisp, as if every plant had been roasted in an oven."[1] Muir first penned these words in 1869, and over the past 150 years California has harnessed its water resources, redistributing the Sierra snowmelt through a labyrinth of federal, state, and regionally financed dams, canal systems, and exchange agreements to become the world's leading agricultural producer. Fresno County, located in the middle of the San Joaquin Valley—that vast area of California from Stockton in the north to the Tehachapi Mountains in the south—is ranked first in the United States for agricultural production, with annual sales of over $3 billion. Vineyards, orchards of almonds, nectarines, and peaches; cotton fields; row crops; and dairies line the rural and state

FIGURES 6.1 and 6.2 (Top) Hmong Strawberry Farmer, Fresno. Photograph by Joel Pickford. (Bottom) Riverview Ranch, Madera County, California, completed 1890; tankhouse, ca. 1900, demolished and rebuilt 1940. Photograph by Karana Hattersley-Drayton.

roads. Small agricultural plots of strawberries and exotic vegetables farmed by Hmong and others of Southeast Asian descent are also an important contribution to the area's agricultural production (fig. 6.1).

The canals built by both the local irrigation districts and the federal Central Valley Project contribute to a distinct regional landscape of great beauty and historic interest. In addition to the water conveyance systems that wind their way through rural and urban areas alike, the San Joaquin

FIGURE 6.3 Kearney Boulevard, Fresno County, California, laid out 1889–92. Photograph by Karana Hattersley-Drayton.

Valley has a historic landscape defined by palm-, olive-, and eucalyptus-lined boulevards and small farms still graced by a tankhouse, transverse crib barn, and foursquare farmhouse or simple bungalow (fig. 6.2).

As preservationists we value this landscape, although protecting it is a challenge. As an example, few water district managers are interested in seeing entire canal systems locally or nationally listed. Fresno, the largest city in the San Joaquin Valley, first adopted a historic preservation ordinance in 1979, which established a local Register of Historic Resources, patterned after the National Register of Historic Places. Fresno's register program is heavily skewed toward buildings, particularly those at the grander end of the scale, with few real examples of working-class housing and no canals or tree-lined boulevards that are so intrinsic to the place. However, one former agricultural colony, two stretches of treescapes, and the fourteen miles of Kearney Boulevard have been evaluated and found eligible for either a county landmark program or the National Register (fig. 6.3). In addition, portions of Central Valley Project canals outside of the Fresno area have been evaluated as eligible to the National Register by the State Department of Transportation staff and their consultants.

Beginning in 2004, the city's Planning and Development Department initiated a Fresno Green Incentive Building Program that was intended to be holistic in its embrace. The program includes standard credits for energy and water efficiency and recycled building content, based on the U.S. Green Building Council's LEED (Leadership in Energy and Environmental Design) program. However, Fresno Green also includes credits specifically tied to the city's 2025 General Plan, as well as points for historic preservation and public art. A developer, for example, can earn one point each for providing a wayside exhibit for a canal, for the inclusion of a historic landscape feature, for conducting a historic survey or for the retention and/or adaptive use of an existing building. The Fresno Green Building Program is incorporated through reference into the citywide Fresno Green Strategy, which additionally calls out the importance of American Public Improvement Standards (APIS) for boulevards with historic landscapes. APIS have been adopted for several streets in the Fresno area and allow for the use of alternatives to the conventional curb and gutter pattern that may detract from the historic feel and character of rural roadways. But how green are these features? How do we balance historic preservation concerns with those of sustainability? Specifically, are canals and historic landscapes really green?

Both canal systems and tree-lined boulevards depend upon the availability of water, and in California water rights have always been politically contentious. Furthermore, the state currently faces a severe water shortage due to several factors: the water supply and delivery system is fragile and outdated; the delta through which more than twenty-five million Californians receive their water is in an ecological crisis; the state's overall water allotment has been reduced through federal efforts to protect endangered species; the population continues to grow, without a concomitant increase in water; and three years of drought (2007–2009) have led to large tracts of land being fallowed in the San Joaquin Valley for lack of irrigation. In addition, recent studies indicate that climate change is reducing the mountain snowpack, which is a critical source of water storage in the state.[2]

The effects of climate change have been documented for the western United States. The Scripps Institution of Oceanography noted that the water content of the western mountain snowpack has increasingly declined since 1950 in eight of the nine mountain regions studied. This decline was definitely linked to climate change. The report concluded that "given the importance of this snowmelt for everything from irrigation to drinking water to powering dams in the West, 'modifications to the water infrastructure of the Western U.S. [is] a virtual necessity'."[3] The negative effects of climate change have also been documented for the San Joaquin Valley in a 2008 study by Fresno State University's Institute of Climate Change Oceans and Atmosphere. The study predicts that there will be a 1-to-2.3 degrees Celsius temperature rise by 2050 with an increase in extreme heat days in the Fresno area. Storms will be fewer but will have greater intensity and the snow level in the Sierra will rise. Water runoff will increase and not be retained as efficiently in the snow pack; the snow will melt earlier in the year leading to increased difficulty of ensuring water delivery for domestic and agricultural use. Crop species will be affected and stone fruits, for example, may not fare well with less water for irrigation and higher temperatures. Pumping of the area's groundwater will be increased, as surface water is decreased and thus the recharge of clean water to the aquifer will be critical.[4]

Fresno is located in a naturally semi-arid region with an average of ten to eleven inches of rainfall a year. It was founded as a railroad town and located ten miles from the San Joaquin River, the nearest major water supply. Water for both domestic and agricultural use was therefore a challenge in the early years although ironically the site had been chosen by Leland Stanford for a new station for the Central Pacific Railroad because he was impressed by the wheat fields of local landowner Anthony Y. Easterby. Easterby and his associate, Moses J. Church, had previously dug one of the area's first canals, which brought water from Fancher Creek. In 1873 the Contract and Finance Company, a subsidiary of the Central Pacific Railroad, filed the plat for Fresno in a gridiron plan that straddled the diagonal rail corridor that cut through the U.S. Rectangular Land Survey of the 1850s.[5]

FIGURE 6.4 Family Posed Next to Irrigation Ditch, Fairmead Colonies, Fresno County. Photograph 1913. Courtesy Julius and Anna Siemens Papers, Center for Mennonite Brethren Studies, Fresno Pacific University.

FIGURES 6.5 and 6.6 (Top) Herndon Canal, looking west toward West Avenue, Fresno. Photograph by Karana Hattersley-Drayton (Bottom) Friant-Kern Canal Off State Route 180, Fresno County. Photograph by Karana Hattersley-Drayton.

Water conveyance systems, along with tree-lined boulevards, were an intrinsic feature of the corporate-sponsored agricultural colonies that were established in and around Fresno, beginning in 1875 with the Central California colony. By 1903 there were forty-eight colonies or tracts in Fresno County representing approximately 71,080 acres. These colonies with their plots of ten-to-forty acres were marketed to farmers around the world and were the primary means by which the San Joaquin Valley was settled. Easterby and Church's Fancher Creek Canal system is now incorporated into the 800 miles of canals and pipelines of the Fresno Irrigation District, which was formed in 1920 under the California Irrigation District Act (fig. 6.4). Six hundred miles of the canals and distribution works were constructed between 1850 and 1880 as part of the privately owned Fresno Canal and Land Company.[6]

Most of these canals, laterals, and ditches remain unlined and open, which creates consequences, both positive and negative, from the perspectives of safety and sustainability. In a region where summer temperatures often exceed one hundred degrees Fahrenheit, canals traditionally serve as the swimming pools for the poor and working classes and are thus a hazard for children due to the swift current and often frigid temperature of the water.

Water loss through evaporation and seepage, or infiltration, is another consequence of an open, unlined canal. Although Fresno's zoning ordinance requires that canals fifty-four inches or less must be piped, most main canals, particularly the Herndon and Dry Creek canals that loop through the city, are too large and thus too expensive to be placed underground (fig. 6.5). Another negative result of an unlined canal is that infiltration can allow chemicals and pollutants carried as runoff to enter adjacent wells as well as the water table.

On the positive side, the replenishment of the aquifer is critical as cities and farmers turn to groundwater pumping. As the authors of the Fresno State University Climate Report caution, "Every opportunity must be made to add to the groundwater table through recharge when water is available." The water table has been increasingly drawn down throughout the San Joaquin Valley with an average drop of 3.2 feet per year.[7] Agricultural colonies that depend upon groundwater for irrigation ultimately failed when the water table fell below the reach of traditional pumps. For example, farmers in Fairmead, north of Fresno, only attained commercially successful yields in the twentieth century after the formation of the Chowchilla Water District in 1948, which tapped into the canal system of the Central Valley Project (fig. 6.6).[8] But water must be clean to be useable and thus the importance of protecting groundwater from city and agricultural runoff that often includes nitrates.

Historically, excess water applied by farmers percolates beyond the root zone and thus recharges the aquifer. As Fresno County has urbanized, less of this water has naturally infiltrated to the water table from agricultural lands. The Fresno Irrigation District has therefore increasingly diverted more of its 500,000 acre-feet of water to groundwater recharge through facilities such as Leaky Acres that includes twenty-six ponds covering 200 acres. The purpose of this facility, constructed in 1970, is to replenish the underlying groundwater by percolating available surface water that enters the site from the Gould Canal. Leaky Acres is ideally sited to be upgradient of the "cone of depression" in the city's sole source aquifer. Highly percolable with sand and no hardpan allows rapid percolation of large quantities to recharge the city's aquifer, which at this point supplies 88 percent of all potable water for Fresno. Unlike storm water basins, which are prone to incidental contamination, Leaky Acres is supplied by relatively unpolluted surface irrigation water. Furthermore, the municipal government has sole jurisdiction over how the site is used and maintained for recharge. The city normally recharges 15,000 acre-feet of water to the aquifer in this concentrated underground pond.[9]

Although unlined and open canals lose water through evaporation and seepage, infiltration also recharges the water table with water that may or may not be of the highest quality. But canal systems offer other features that need to be considered in evaluating their contributions to a sustainable community, including aesthetics, a potential for recreation, and their importance in defining regional history and sense of place. Unlike the former industrial canals of Britain and Europe that were constructed to accommodate shipping on at least a modest scale, these agricultural canals have a series of drop structures that prevent their ready use for boating and excursions. They are also still working canals with water released seasonably as needed. Water delivery for the Fresno Irrigation District in 2008 started on the first of March and concluded on the thirty-first of July due to the fact that the flow in the Kings River was only 71.5 percent of normal. At the end of the season the canals are "de-watered" revealing castoff tires, shopping carts, and other debris. When flowing, however, the miles of canals offer recreational opportunities for walking and jogging. Unfortunately, plans to develop an official Bankside Trail project along the Herndon Canal has been sidetracked by water district concerns over liability and access.

The dry, stark landscape of the mid-nineteenth century has, through water, been transformed into not only the richest agricultural region in the world but also what boosters often described as a Garden in the Sun. Lush environments designed by Thomas Church, Rudolph Ulrich, and other landscape architects are the proud hallmark of Fresno neighborhoods such as Fig Garden and Sunnyside. Unfortunately most of the landscapes are not drought-tolerant and consume approximately 70 percent of residential water. Fresno houses do not have water meters and perhaps as a consequence the city has one of the highest water uses in the southwestern Sun Belt at 294 gallons per capita each day. Water meters will be required by 2013 and it is expected that reduction of household irrigation will be "an easy way to conserve water." The drying out of vegetation in the city could, however, lead to a higher incidence of urban wildfires.[10]

One bright note is that the treescapes planted by pioneers, such as the fourteen miles of Kearney Boulevard, are lined with species that are naturally drought-resistant: mission olives, Eucalyptus viminalis (mama gum from Australia), Nerium oleander (orealnder), and Washingtonia filfifer (California fan palm) require no irrigation after the first few years.[11] Unfortunately these trees are often in the path of change and several stretches have been removed for public works projects. One current city project, the widening of Peach Avenue, will take out the one-hundred-year-old allée of mission olive trees. However recent mitigation measures will preserve in place as many of these trees as possible; in addition, non-fruit-bearing olives will be used in the planting strips along the widened road. In another example, developers along the historic Kearney Boulevard were required to prepare a planting pattern as none existed from the late nineteenth century. Dead and dying trees will be replaced with new, hardier varieties.

How "green" are the historic landscapes that we value as preservationists? Is brown the new green, or not? Unlined canals are both water wasters and water savers in that water lost through seepage re-enters the region's water table, from which most urban water is drawn. Canals form the aesthetic backbone of many communities and are an important visual link to the region's settlement when most agricultural colonies were both irrigated and criss-crossed by tree-lined boulevards. The potential for more active recreational use on local canals should also be explored.

For gardeners, as water becomes ever more precious, local agencies will want to consider cash-for-grass programs, such as instituted in Las Vegas, and will need to educate the public more actively about drought-tolerant plants. The City of Fresno is currently implementing the state's Model Water Efficient Landscape Ordinance enacted on 1 January 2010, which sets standards for landscaping in new developments. Thankfully, the pioneers who settled the region were educated horticulturists and they chose to plant the early boulevards with trees such as eucalyptus, olives, and palms that will do well in the hotter summers that are predicted for the future. ❁

Notes

1 John Muir, *My First Summer in the Sierra*, 1911; reprint ed. (San Francisco: Sierra Club Books, 1988), 1.

2 "California's Water: A Crisis We Can't Ignore," Association of California Water Agencies, September 2007; see *http://www.acwa.com*.

3 Thomas L. Friedman, *Hot, Flat and Crowded: Why We Need a Green Revolution—And How It Can Renew America* (New York: Farrar, Straus and Giroux, 2008), 132.

4 Fraka Harmsen, et al., "Mitigation and Adaptation Strategies for Climate Change in Fresno California," Institute of Climate Change Oceans and Atmosphere, College of Science and Mathematics, California State University, Fresno, August 2008.

5 Charles W. Clough and William B. Secrest, Jr., *Fresno County, The Pioneer Years: From the Beginnings to 1900* (Fresno, Calif.: Panorama West Books, 1984), 117-18; John W. Reps, *Cities of the American West: A History of Frontier Urban Planning* (Princeton: Princeton University Press, 1979), 184-87.

6 John Panter, "Central California Colony: 'Marvel of the Desert'," *Fresno Past and Present* 36 (summer 1994): 9; see *http://www.fresnoirrigation.com/history*.

7 Harmsen, "Migration and Adaptation," 51–52; personal communication from Joseph Oldham, sustainability manager, City of Fresno, 4 August 2008.

8 Stephen D. Mikesell, "The Central Valley Water Project: Vernacular Architecture Writ Large," in Karana Hattersley-Drayton, ed., "Architecture, Ethnicity, and Historic Landscapes of California's San Joaquin Valley," Planning and Development Department, City of Fresno, April 2008, 21.

9 Personal communications with Brock Buche, supervising engineer for Leaky Acres, and Sandra Brock, planner, City of Fresno, 5 March 2010.

10 Harmsen, "Migration and Adaptation, 53.

11 Personal communication with Hilary Kimber, parks supervisor, City of Fresno, 28 January 2009.

STEWARDSHIP AND SUSTAINABILITY OF HISTORIC WATERWORKS INFRASTRUCTURE

Meisha Hunter

The history of civilization is the history of infrastructure.
Oliver Schwaner-Albright [1]

Historic waterworks infrastructure should be preserved, whether operable or decommissioned, whether in whole or in part, whether above ground or below, because they inform modern water managers, policy makers, preservationists, and designers of the heroic civil engineering achievements of the past realized in the face of needed social change. Wherever they were constructed—whether in ancient Rome or industrial-era New York—aqueducts were created to deliver adequate and reliable supplies of potable water to consumers, and their presence in an urban community consistently reduced outbreaks of disease and fire. Amidst a culture of technological invention and infrastructural expansion, the survival of this often unseen and silent utilitarian tissue has become increasingly vulnerable, in some cases threatening the future of communities that rely upon them. The widening gulf between a fragile past and modern infrastructure management practices should galvanize debate about the importance of hydraulic heritage within the preservation community, since the field's purview is the protection of the built environment by physical and intellectual means.

The gap between recognizing the intrinsic engineering, scientific, aesthetic, and social values of historic waterworks and allowing that heritage to perform a didactic, public-service function amidst the very real challenges of managing water systems as cultural heritage is not easily bridged. It is also particularly difficult to value and protect heritage that is physically concealed, publicly inaccessible, and silent by design. While this heritage is increasingly benefiting from designation and recognition, it is not benefiting from a focused discussion about what it teaches us, why it should be preserved, and how it should be managed. Absent informed decision-making, which is the primary task of historic preservation, this heritage can face unsympathetic alteration and encroachment, obsolescence, abandonment, and demolition. Amidst a recent presidential initiative to realize infrastructure renewal and expansion as a national priority, there is a tremendous need to expand the existing preservation canon regarding viable models for sustainable stewardship management of hydraulic infrastructural heritage, and to actively engage the water management, engineering, planning, and policy-making communities in debate about this critical subject (fig. 7.1).

FIGURE 7.1 City Tunnel #3, interior view. Courtesy New York City Department of Environmental Protection Archives.

Historical Overview

Unlike many modern engineers who privilege new designs and technology, some civil engineers of the past observed preservation opportunities when confronted with abandoned waterworks infrastructure. Specifically, these engineers envisioned rehabilitation programs that preserved as much historic fabric as possible while restoring infrastructure to its purpose-built use. For example, after surveying the ancient Roman Pont du Gard aqueduct bridge in Nîmes, France, civil engineer Fayette B. Tower (1817–57) wrote in his *Illustrations of the Croton Aqueduct* that "this structure, which has been out of use during fourteen hundred years, is still in such a state of preservation that it could be restored without a very great expenditure of money" (fig. 7.2).[2] When Tower contemplated the ancient bridge, he did not just see a monumental masterwork in stone that had survived centuries of decay. He recognized rehabilitation as pragmatic and economical with the goal of returning a unused, but valuable, resource to public service. Tower's vision—that even ancient waterworks heritage can be an asset waiting to be realized—should inspire modern engineers to reassess the long-term utility and public service potential of hydraulic infrastructure.

FIGURE 7.2 "Brucke über den Gard" (Pont du Gard), from *Museum des Wundervollen, oder Magazin des Ausserordentlichen in der Natur, der Kunst und im Menschenleben* (1803–5). Picture Collection, New York Public Library, Astor, Lenox and Tilden Foundations.

The preservation, engineering, water management, and policy-making communities need to acknowledge the vast portfolio of waterworks infrastructure that already exists and the enormous investment in embodied energy that this heritage represents. Whether constructed to meet propagandistic, strategic, or civic-virtue purposes, monumental waterworks systems such as the Aquedotto Vergine (dedicated 19 B.C.E.) in Rome and the Croton Aqueducts (Old Croton 1837–42; New Croton 1885–90) in New York were built of durable materials according to engineering designs that anticipated their long-term serviceability. Consider that in Rome, an ancient aqueduct continues to function, while in New York, a mid-nineteenth-century

FIGURE 7.3 High Bridge, New York, 1837–42, John B. Jervis, engineer, James Renwick, Jr., associate architect; center arches replaced by steel structure 1927. Photograph 1916. Courtesy New York City Department of Environmental Protection Archives.

aqueduct was decommissioned after only 123 years of use. The comparison between the two aqueducts' stewardship bears closer analysis. Within the same period, from 1955 to 1965, as New York's Department of Environmental Protection (DEP) decided to de-activate a forty-one-mile-long, gravity-fed, masonry-lined aqueduct that had been brought on-line in 1842 (fig. 7.3), while Rome's Azienda Comunale Energia e Ambiente (ACEA) surveyed and ultimately chose to repurpose a 2000-year-old, twenty-one-kilometer-long, gravity-fed, masonry-lined aqueduct that was deteriorated but still active (figs. 7.4-7.5). During the post-World War II period in Rome, unregulated residential construction above the Vergine—and the associated absence of wastewater utilities—resulted in soil contamination, environmental degradation, infiltration, and structural damage that necessitated declassification of the aqueduct's water from potable to non-potable status.[3] Instead of abandoning or demolishing the aqueduct, ACEA undertook a structural reinforcement campaign and repurposed the aqueduct's water for ornamental fountain displays and the irrigation of public parks (flow rate=300 ltr/sec).[4] Such startling contrasts between the long-term utility of these two aqueducts challenges not only preservationists but also engineers, planners, and policy makers concerned with developing an inclusive, sustainable future that promotes holistic preservation management of our nation's infrastructure.

FIGURES 7.4 and 7.5 (Left) Vergine's channel under the Tombino di Via Cave di Pietralate, cross-section. Courtesy ACEA Spa. (Right) Aquedotto Vergine, Chocciola, Pincian Hill, Rome, interior view. Courtesy Allan Ceen—Stadium Urbis.

As demonstrated by the Vergine, a remarkable and recurring fact about water systems is that age does not necessarily negate serviceability or utility.[5] In a global culture that increasingly seeks to espouse the maxim of "reduce, reuse, recycle," the reality of historic, functional waterworks is valuable to consider. After decades of abandonment, a re-activated, industrial-era masonry conduit delivers gravity-fed drinking water to a community in Westchester County. Etruscan drainage canals and subterranean tunnels, hand-carved from volcanic tufa over 2,800 years ago, continue to operate in agricultural fields north of Rome. Hand-drawn water from ancient springs in Bulla Regia provides an economically viable alternative to modern private concessions for agrarian communities in rural Tunisia. Rainwater collected in underground cisterns supplement precious desalinated drinking-water supplies in Malta and the Caribbean. And nineteenth-century masonry culverts in Westchester County continue to divert streams from the extant aqueduct channel above.

Amidst increasing public consciousness about non-renewable resources and the carbon footprint legacy of demolition, it is important to consider that antiquated waterworks can be re-purposed after they are decommissioned. Stewardship of waterworks can apply to active as well as de-activated elements. Environmentally sustainable, economically viable, and architecturally feasible alternatives to abandonment and demolition are possible. Adaptive use has transformed reservoirs into swimming pools (Highbridge Recreation Center in New York), pressure equalizing towers into lookouts (Highbridge Tower), wells and waste weirs into publicly accessible sites (Pozzo di San

Patrizio in Orvieto, Italy; Ossining Waste Weir in Westchester County, New York), gatehouses into performing arts venues (135th Street Gatehouse/Harlem Stage) and adult day care centers (113th Street Gatehouse), cisterns into concert halls (Palace Cistern in Istanbul), and ventilators into mile markers on scenic trails (Old Croton Aqueduct State Park in Westchester County).

These inspiring examples parallel a growing momentum within the field of historic preservation to consider waterworks systems, whether in whole or in part, as industrial heritage and cultural patrimony. This approach is consistent with the International Committee for the Conservation of the Industrial Heritage's draft, Nizhny Tagil Charter for the Industrial Heritage.[6] The charter was adopted by the committee at its XIIth Congress in Russia in 2003, and is the international standard for the study, documentation, conservation, and interpretation of industrial heritage. In addition to the charter, it is possible to argue that the twentieth century has evidenced concerted efforts towards infrastructure documentation, recognition, and historic designation. As of February 2008, 543 properties with historic waterworks sub-functions have been listed on the National Register of Historic Places (0.6 percent of 83,000 properties nationwide). These encouraging statistics bear witness to the contributions of countless individuals and teams, scholars, engineers, architectural historians, and preservation professionals whose cumulative efforts have made these nominations possible. However, historic designation of waterworks heritage must be complemented by preservation management in order to balance authenticity and integrity of historic fabric with technological upgrade and expansion campaigns.

In the past, this heritage was frequently overlooked, whether because much of it was physically concealed and therefore more difficult to appreciate than the visually prominent architectural heritage, or because the majority of officially recognized heritage in the United States is defined as architectural rather than engineered. As the economist Kate Ascher argues in *The Works: Anatomy of a City*: "Rarely does a resident of any of the world's great metropolitan areas pause to consider … the myriad systems that operate round the clock to support it … even the most mundane domestic tasks would be impossible without the far-reaching, complex, and often invisible networks of infrastructure that supports them."[7]

Closed-Loop Resource Management

In order to safeguard the embodied energy of historic waterworks infrastructure and maximize its public service utility for future generations, sustainable stewardship must be considered applicable to active and decommissioned components alike. Acknowledging waterworks infrastructure as a renewable, rather than a disposable, resource is consistent with "closed-loop" resource management. A 2006 conference paper, "Sustainable Infrastructure: Doing More with Less," presented before the United Nations Economic and Social Commission for Asia and the Pacific, argues in favor of applying the maxim of "reduce, reuse, recycle" to infrastructure. The author emphasized two important points: first, that shared use of infrastructure and adaptively programming of underutilized or surplus infrastructure before investing in new construction yields tangible benefits and, second, that applying a holistic, system-wide approach can result in more effective use of existing resources.

Arguably, waterworks infrastructure that maintains its purpose-built function may have the greatest possibility of retaining character-defining features, historic fabric, and physical integrity. More than three decades after the DEP's decision to decommission the Old Croton Aqueduct, an abandoned three-and-one-half-mile segment of conduit was re-activated in 1986 to serve the village of Ossining in Westchester County.[8] Benefiting from the only instance of Croton water delivery via the Old Croton conduit today, 30,000 residents receive 63 percent of their drinking water from this source.[9] The DEP is also committed to rehabilitating and improving the New Croton Dam, which is designated a national historic landmark. The project, which seeks to extend the service life of the facility by fifty-to-one-hundred years, will include preserving the historic dam and its associated structures, performing masonry repairs and cleaning, and rehabilitating water conveyance systems. Lastly, the DEP has recognized the value of historic infrastructure in terms of creating redundancy within the system. The anticipated opening of City Tunnel #3 in 2012 will allow for temporary closures and inspections of the two existing tunnels (1917 and 1936), which have not been closed in over seventy years (see fig. 7.1).[10] These examples underscore the fact that abandoned nineteenth-century infrastructure can be reactivated and that the service life of that infrastructure can be extended and strategically integrated to assist modern communities in meeting their drinking water needs.

Historic waterworks systems that were constructed before the advent of electrically generated power also benefit from energy efficiencies in delivery. Designed to operate twenty-four hours a day all year, and relying on gravity to deliver water from source to terminus, these systems are inherently more energy efficient than a system that depends on mechanization. Aside from a 5 percent incidence rate of drought, New York's water supply is 95 percent gravity-fed, necessitating a low dependence on and consumption of energy to meet demand. Arguably, sustainable stewardship of pre-industrial infrastructure may become increasingly relevant within global conversations on emissions reduction and climate change.

Repurposing and rehabilitation of formerly threatened monumental industrial heritage is gaining ground in New York City. The adaptive use and reprogramming of the High Line in Manhattan as an elevated pedestrian walkway and linear park, and the city's Department of Parks and Recreation's recent RFP for rehabilitating the High Bridge to reconnect Manhattan and Bronx boroughs are inspiring and possibly momentum-building initiatives. Mayor Michael Bloomberg marked the completion of the $4.2 million renovation project linking High Bridge Park and High Bridge by saying, "I'd like to think that the High Bridge will do for uptown and the Bronx what the High Line is going to do to revitalize the neighborhoods along the Hudson River."[11] The reactivation of industrial heritage within New York's stratified urban landscape and the economic regeneration of surrounding neighborhoods that these projects could trigger may encourage policy makers, economists, and planners to examine this phenomenon more closely.

Making the Case for Preservation

In the past nine years, the Environmental Protection Agency, the Congressional Budget Office, the Water Infrastructure Network, and the Harvard School of Public Health have prepared independent studies criticizing inadequacies in funding for maintenance and repair of public water infrastructure over the past half-century and have forecasted capital investments needed for the future maintenance and expansion of the nation's water and wastewater infrastructure.[12] In 2005, a nationwide survey of infrastructure in the United States conducted by the American Society of Civil Engineers (ASCE) accorded a "D-" rating for water infrastructure and called for a $1.6-trillion-dollar investment over a five-year period to address deficient conditions.[13] The ASCE urged for the establishment of a long-term development and maintenance plan as a national priority and challenged Congress to increase federal water infrastructure spending to help bridge the current $11 billion annual shortfall. The importance of the ASCE's determination and corresponding remediation-cost forecasting remains clear: if left unchecked, the nation's infrastructure, and by association the population at large, will face an epic crisis. It is not so much a question of whether to address these problems, but how.

Without sustained capital investment, comprehensive management, and implementation of prudent stewardship strategies affecting water resources and infrastructure, these critical utility systems could fail. Despite the fact that survey, cyclical maintenance, and sympathetic repair are necessary to continue the operable service life of infrastructure, the tools, materials, methods, agency resources, and financial capital required to do so may not exist. Historic infrastructure may run counter to a prevailing mind-set among some water managers and civil engineers, who believe that "new is better."[14] Political willingness to make stewardship and sustainability a priority may be absent. Roy Sparrow, professor emeritus of Public Management at New York University's Wagner School of Public Service, has posited that committing government dollars to improvements that may be invisible to the public makes spending on infrastructure politically unpopular. During an interview with a reporter from the *New York Times* about the city's crumbling infrastructure, Sparrow said, "'I can't recall any politician who ran successfully on the issue of infrastructure."[15] In this period of evaluating feasible, cost-effective pathways to address the nation's impending infrastructure crisis, closely considering stewardship and sustainability, and reexamining the relationship of historic infrastructure to the built environment will be timely. Maintaining and rehabilitating historic infrastructure wherever possible to promote purpose-built utility and repurposing decommissioned infrastructure will demonstrate sustainable stewardship in action.

Sustainable Stewardship Models

Sustainable stewardship can be as basic and necessary as recycling. Applied to waterworks infrastructure, the challenge to promote the operability of existing infrastructure or the repurposing of decommissioned infrastructure must be met by preservationists, water managers, policy makers, and planners. For active waterworks with engineering, architectural, or historical significance, a set of systematic standards for maintenance and conservation, inclusive of water management and preservation concerns, must be codified and adopted. Measures that facilitate jurisdictional transfer to and repurposing of discrete elements should be considered as systems continue to age and as elements that are functional today may be decommissioned in the future.

Two preservation management models, affecting still-functioning historic bridges and decommissioned lighthouses, merit closer study. In 2004, the American Association of State Highway and Transportation Officials' Standing Committee on the Environment requested the preparation of the "Guidelines for Historic Bridge Rehabilitation and Replacement."[16] The guidelines, which were endorsed and published by the committee in 2008, are a nationwide best-practices standard that outlines when rehabilitation can be considered prudent and feasible. Arguably, the dissemination of these guidelines within the historic waterworks community—including water managers, policy makers, and preservation agencies—could stimulate substantive debate about the prudency of rehabilitation versus replacement of historic waterworks infrastructure.

In 2000, the National Historic Lighthouse Preservation Act was passed, establishing a protocol for the disposal and post-deactivation stewardship of federally owned, National Register-listed or -eligible lighthouses. The collaborative government program coordinates the efforts of the U.S. Coast Guard, the General Services Administration, the National Park Service, the Park Service's Maritime History Program, and respective state historic preservation offices in sharing documentation, Section 106 review, and title transfer. As of June 2007, of the seventy-four lighthouses offered under the provisions of the act, twenty-five have been transferred to new owners—including other federal agencies, state and local governments, and non-profit organizations—and eight have been auctioned by the GSA.[17]

What remains to be more fully developed within the preservation realm is a body of literature that specifically addresses the unique concerns of balancing water management needs and preservation concerns and offers models for preservation management of such heritage within active supply systems. One such document is the "Memorandum of Agreement Concerning the Continued Operation of Jerome Park Reservoir after Listing to the New York State and National Registers of Historic Places" (MOA), executed by the New York State Office of Parks, Recreation and Historic Preservation; the New York State Historic Preservation Office (SHPO); and the DEP on 11 July 2000. Originally constructed between 1895 and 1906 at the southern confluence of the Old and New Croton aqueducts, the 500-million-gallon reservoir currently functions as a balancing reservoir for incoming water collected from the upstate Croton watershed and its outgoing gravity-fed supply is delivered both to the Bronx and Manhattan.

After protracted discussions between the SHPO and the DEP concerning the consequences of listing an operable reservoir as a historic resource within the water supply system and the DEP's ability to meet security standards and water management needs, the site was listed on both the state and national registers in 2000. The MOA, which represents a working agreement between three agencies over the future governance of the reservoir, evidences a spirit cooperation and pragmatism, recognizes the DEP's current and future needs to "rehabilitate, modify, upgrade, expand and/or make additions" to the reservoir, and respects SHPO's determination of the site's multivalent significance.

FIGURES 7.6 and 7.7 (Top) Falkirk Interchange, near Falkirk, Scotland, 1999–2001, Ove Arup & Partners, engineers. General View. Courtesy British Waterways Scotland. (Bottom) Falkirk Interchange, detail of wheel. Courtesy British Waterways Scotland.

Future Currents

Management of historic waterworks will, of necessity, continue to demand alteration, upgrade, and expansion. Collaborations between engineers, planners, water managers, and preservationists, who unite sustainable design and cultural heritage, have already produced exciting results. One example is that of British Waterway's Millennium Link, an ambitious canal restoration project in central Scotland, which ultimately reconnected the historic Forth & Clyde (begun 1768) and Union (1818-22) canals between Glasgow and Edinburgh.[18] The two waterways had been linked by a series of eleven locks, which had been dismantled in 1933. Completed in 2001, the project's centerpiece, a € 78 million Falkirk Interchange designed by Ove Arup & Partners, features a 115-feet-high-by-115-feet-wide-by-100-feet-long wheel designed with a 120-year service life, which can lift up to twenty boats (6,000 tons) between an upper and lower pond at the juncture of the two canals (figs. 7.6, 7.7). Ultimately, such a stunning model of sustainable stewardship, which boldly combines cutting-edge technology and design with extant but formerly disconnected hydraulic infrastructure, should inspire water managers, engineers, policy makers, and preservationists throughout the world for years to come.

Notes

1 Oliver Schwaner-Albright, "Design Challenge: Fend Off Disaster," *New York Sun*, 22 July 2008.

2 Historic preservation: "Encompasses a broad range of activities related to preservation and conservation of the built environment by physical and intellectual methods." Ward Bucher, *Dictionary of Building Preservation* (New York: Wiley, 1996), 235.

3 W. Brown Morton, III, "The Secretary of the Interior's Standards for Historic Preservation Protects: Ethics in Action," paper given at the Annual Meeting of the National Council for Preservation Education, Indianapolis, 23 Oct. 1993, 15–23, copy in author's possession.

4 Fayette B. Tower, *Illustrations of the Croton Aqueduct* (New York: Wiley and Putnam, 1843), 20.

5 This undated, internal document was provided by ACEA staff at the Serbatoio del Gianicolo to the author. Regarding illegal settlements: in 1975 ACEA was entrusted with the task of expanding service to illegal settlements in peripheral urban areas, or borgate. A decade later it became responsible for operating wastewater treatment. Extension of service to the borgate, which ACEA previously served by tank, was an important contribution to Rome's sustainable development. Around 350,000 people live in eighty-two borgate, accounting for twelve percent of the city's population and raw wastewater used to contaminate groundwater. See *Watertime: Case Study for Rome* (Brussels: European Commission; and London: PSIRU School of Business, University of Greenwich, 2005), 6. For a discussion of the phenomenon of condo, or Rome's forgiveness of

violators of municipal and state codes, including illegal construction, see Anthony Tung, *Preserving the World's Great Cities* (New York: Three Rivers Press, 2001), 66.

6 Vittorio Nicolazzo, *L'Acqua Vergine: E Suoi Acquedotti e le Fontane di Roma Attraverso I Secoli* (Rome: ACEA, 1998), 207, and an undated internal ACEA document made available to the author.

7 The nineteenth-century American engineer Fayette B. Tower, who worked as an assistant engineer on the first Croton Aqueduct campaign in New York, prepared a narrative of its construction in 1843. In the preface to *Illustrations of the Croton Aqueduct*, he surveyed ancient aqueduct systems, particularly the Pont du Gard.

8 For the full text, see *http://www.international.icomos.org/18thapril/2006/nizhny-tagil-charter-e.pdf*.

9 A similar observation was made by a Chadian immigrant to New York: "I sat marveling at network ... constructed by builders, plumbers, and electricians, each trained in their craft. I looked out into New York streets and contemplated what it takes to give a city water and electricity 24 hours a day, regular trash pickup, subways, phones and the internet. The answer: huge numbers of people, systems it took years to create and overlapping technologies all conspire to bring us a life that many take for granted in the developed world." See *http://www.psdblog.worldbank.org/psdblog/2006/11/life_interrupted*.

10 "Village of Ossining Comprehensive Plan," Village of Ossining Comprehensive Plan Steering Committee with Philips, Preiss, Shapiro Associates, 2007, n.p.

11 "Annual Drinking Water Quality Report," Village of Ossining, 2006; see *http://www.villageofossining.org/documents/aqwr_2006.pdf*.

12 See *http://www.nyc.gov/html/planyc2030/downloads/pdf/report_water_network.pdf*.

13 "Highbridge Park Path Reopened," *Gothamist*, 27 August 2008, n.p.

14 Ronnie B. Levin, et al., "U.S. Drinking Water Challenges in the Twenty-First Century," Harvard School of Public Health, *Environmental Health Perspectives*, Supplement Volume 110, No. S1, 2002.

15 "Report Card for American Infrastructure," American Society of Civil Engineers, 2005; see *http://www.asce.org/reportcard/2005/index2005.cfm*.

16 The observations of architect and preservationist, James Marston Fitch, support this hypothesis: "The general tendency of ... dam and highway engineers, has been to dismiss as romantic or sentimental the efforts to protect historical impediments to their special brand of "progress." He also argued that "change is now seen as synonymous with progress." James Marston Fitch, *Historic Preservation: Cultural Heritage Management of the Built World* (Charlottesville: University Press of Virginia, 1990), 23, 31.

17 John Sullivan, "Study Puts High Price on Fixing A Crumbling City," *New York Times*, 25 August 1998.

18 Patrick J. Harshbarger, et al., "Guidelines for Historic Bridge Rehabilitation and Replacement," Lichtenstein Consulting Engineers in association with Parsons Brinckerhoff Quade & Douglas, Paramus, New Jersey, 2007. The executive summary also describes some concerns and questions that may apply to historic waterworks: "Research indicates that while many transportation agencies have processes for managing historic bridges, including some with well thought out treatments for preserving specific bridges, few have written protocols or guidelines that lead engineers and historians through the decision-making process. When is it not prudent to rehabilitate a substandard bridge? When is it prudent to upgrade the bridge to keep it in service? What work can be done without adversely affecting what makes the bridge historic? Has everything that could be done to preserve the bridge been considered?

Research indicates that engineers believe that their agencies are proactive about historic bridge preservation while historians do not. So ... the perception is that historic bridges that could be saved are not being saved because rehabilitation is not as consistently considered as it should be."

19 See *http://www.nps.gov/history/maritime/nhlpa/nhlpa.htm*.

20 The author is indebted to civil engineer Craig Covil of Arup's New York office for highlighting the Falkirk Interchange case; see *http://www.arup.com/maritime/project.cfm?pageid2281*.

Sustainability and Preservation of Affordable Housing: Greening Row Houses in Philadelphia

Liz Robinson

In 1984, Philadelphia's energy-poverty crisis reached an unprecedented level, with more than 30,000 residential gas service customers having their service terminated for non-payment. The municipal government recognized the need to develop a system that would effectively coordinate all low-income energy services and awarded a small grant for the Energy Poverty Task Force to found the Energy Coordinating Agency (ECA), an entrepreneurial, non-profit organization. Since its establishment, ECA has grown steadily from a staff of four to 125. The organization's mission is to help people conserve energy and to promote a sustainable and socially equitable energy future for all residents of the metropolitan area. ECA administers almost all the low-income energy services in the city, encompassing energy conservation measures, heating system repair and replacement, bill payment and assistance, and energy education services to more than 40,000 low-income households. Its energy conservation programs prevent the emission of over 2,500 tons of carbon dioxide annually. ECA is one of six weatherization training centers under contract to the Pennsylvania Department of Labor and Industry to provide comprehensive weatherization training. As a Building Performance Institute affiliate, ECA is training and certifying building analysts throughout the Mid-Atlantic region. ECA is currently completing renovation of its Knights Green Training Center, converting a 56,000-square-foot factory into a LEED Gold-certified building-science training center, which will embody the sustainability technologies and principles that are being taught there. This facility will have state-of-the-art classrooms and laboratories to prepare men and women for a wide range of jobs in energy-related industries.[1]

ECA has also become a leader in the development of energy efficient, affordable, and market-rate housing, sustainable building practice, and solar energy in Pennsylvania. As a LEED for Homes Provider, ECA works with a wide range of developers, builders, architects, non-profit organizations, and others to bring new houses to the Energy Star Homes and LEED for Homes standards. LEED for Homes is the leading standard for sustainability in housing and is supported by the U.S. Green Building Council. ECA received the 2008 Energy Star Leadership in Housing Award from the U. S. Environmental Protection Agency for its work in affordable housing. With the support of the Clinton Climate initiative, BASF, Dow Chemical, and many other corporations, ECA is partnering with the City of Philadelphia and the Commonwealth of Pennsylvania to implement their sustainability plans. Mayor Michael Nutter has called for Philadelphia to become the "greenest city in America" and has laid out a very ambitious sustainability plan, "Greenworks Philadelphia."[2]

Traditional wisdom has it that new houses are far more energy efficient than existing ones. That belief is not entirely true. It is easier, of course, to build energy-efficient houses than to correct the myriad problems inherent with existing ones, and it is essential that all new houses be very energy efficient. But given the embodied energy in existing buildings, the loss of that energy if they are demolished, and the energy expended on demolition and removal, coupled with the fact that 20 percent of our energy use comes from the residential sector, it is even more important that energy use be reduced in existing houses.[3] New construction represents a relatively small and declining share of the market, so it is essential to focus on making existing houses more affordable and on lowering their carbon footprint. Moreover, green residential construction may have its own shortcomings. Recent evaluations have found, for example, that in New York State new Energy Star Homes in fact use more energy than the average existing residences.[4] Why? Primarily because the new buildings are larger and because the Energy Star Homes standard is not that far above code.

The rising cost of almost all building materials, combined with a very soft housing market and the increased importance of sustainability, have caused many city planners and policy makers to look more favorably on housing preservation than new construction in both the affordable and market-rate sectors. Policy, particularly at the municipal level, is beginning to shift away from new construction to housing and community preservation. Leaders in historic cities such as Philadelphia are finally realizing the incredible value of their existing buildings. Even the humble brick row house is gaining in favor.[5] In fact, more people are recognizing that the row house is an inherently efficient structure: it is compact, has a relatively small footprint, and, with its party walls, its heating and cooling loads are inherently smaller (figs. 8.1, 8.2).

FIGURES 8.1 and 8.2 (Top) Cool block in Bridesburg neighborhood, Philadelphia. Photograph by ECA staff. Courtesy of ECA. (Bottom) Typical row houses for blue-collar families from around the mid-nineteenth century, Philadelphia. Photograph by ECA staff. Courtesy of ECA.

Increasing energy prices, the mortgage foreclosure crisis, and our current economic woes seem to have made many more policy makers aware of the need for affordability in housing—not just the purchase price, but the operating costs. Energy efficiency for existing houses is back in fashion! Due to the swift rise in energy prices over the last several years and the growing concerns over climate change and carbon emissions, there is also a rapidly growing movement toward deeper energy savings in existing houses. In the past, programs were content to achieve an average of 20 percent energy savings. Now, an increasing number of people are aiming for 40 or 50 percent savings or even higher. There is a growing interest in achieving net zero energy, even for existing houses. A net zero energy house is one that is extremely energy efficient and uses on-site renewables, usually solar power, to produce what energy it needs. In Europe, this standard is commonly known as Passivhaus. This situation creates a real challenge and opportunity for historic preservation practice to incorporate much higher levels of energy efficiency in order to reduce the operating cost of historic sites and houses, and therefore to increase the likelihood of their preservation.

The EPA's energy conservation program for existing houses, Home Performance with Energy Star, is now mature, with a proven track record capable of responsibly handling the energy efficiency retrofitting of historic buildings.[6] Home Performance programs aim for 20 percent savings on average. A number of new energy efficiency technologies and materials have the ability to help preserve existing buildings and to match materials in historic preservation retrofits.

One such material is cool roof coating, which can be tinted to match any color and can restore old metal roofing such as standing seam roofs. In Philadelphia, ECA began using white roof coatings in an effort to develop an affordable cooling approach for houses occupied by elderly, low-income people. The 1990s was the hottest decade on record. During that ten-year period, Philadelphia experienced four summers with "killer" heat waves. During the summer of 1999 alone, 164 people died in their own homes as a result of these extreme levels in temperature.[7] Most were elderly, low-income people, living without air conditioning or with a single window unit in one room. The Philadelphia row house is essentially a brick box with a flat black tar roof. During the summer, these roofs become unintentional solar collectors, reaching temperatures as high as 198 degrees Fahrenheit, and transferring much of that heat to the brick walls and other materials of the house. On the third day of a heat wave, it is not uncommon for the indoor temperature in an un-air-conditioned house to reach 115 degrees on the top floor, even though the temperature outside may only be 98 degrees.[8] If the windows are closed and the elderly occupants no longer have the strength in their arms to lift them, the temperature on the top floor of the house under that black roof keeps building day after blistering day. The house literally becomes a brick oven. As global warming proceeds, these killer heat waves are predicted to become more frequent and more intense.[9]

In the past the federal government has responded to this crisis by distributing room air conditioners and fans after the fact, and by giving a one-time grant of $100 on the electric bill to low-income, elderly households. The unhappy experience of administering this program in Philadelphia motivated ECA to resolve to develop a better response to the problem of increasingly frequent and deadly heat waves. ECA designed the Cool Homes Program, and with funding from local foundations, corporations, and eventually the U.S. Department of Health and Human Services, ECA has thus far changed almost 600 roofs in the city from black to white.

The key treatment in this affordable cooling program is a white, reflective roof (figs. 8.3, 8.4). Our preferred material for an existing roof is acrylic, elastomeric roof coating. This coating has a reflectance of approximately 92 percent when it is applied, and it has very high emissivity, which allows it to release its heat to the night sky as soon as the sun goes down each day. Elastometric coatings reduce air-conditioning load by approximately 22 percent in the Mid-Atlantic region.[10] An independent evaluation of this pilot program found that for a house without air conditioning, there was a four-degree Fahrenheit reduction in temperature on the top floor of the house once the coating had been applied.[11] Elastomeric coatings have a number

FIGURES 8.3 and 8.4 (Left) Contrast between elastometric and conventional tar and asphalt coatings, Philadelphia. Photograph by ECA staff. Courtesy of ECA. (Right) ECA staff applying elastometric coating over existing roof, Philadelphia. Photograph by ECA staff. Courtesy ECA.

of properties that make them ideal for use in historic preservation. To be reflective, they do not have to be white. In fact, they can be pigmented to match almost any color. The greenish blue of oxidized copper, or a dull red traditional to other metal roofs, are easily matched in these coatings. When the coatings are pigmented, they lose only a little of their reflectance. Together with other related materials, these reflective roof coatings can be effectively used to restore a metal or asphalt roof in poor condition and to preserve the underlying roof by an average of ten years. If properly applied, elastomeric roof coatings should last at least ten years, and will simply need to be recoated at that time.

When a number of roofs concentrated in a relatively small area are coated, the benefits are even greater. ECA started doing roof coatings through a scattered-site approach, one house at a time, but quickly moved to the "cool block" approach, which is more efficient and effective. Thus far ECA has done three cool blocks in Philadelphia, in which it provided energy conservation treatments, including white roof coating, roof insulation, and other energy-saving measures to every dwelling on the block (figs. 8.5, 8.6).[12] Because row houses are connected to their neighbors, the block is really one structure. If the block has sixty row houses, treating all of them will yield greater benefits than treating sixty individual dwellings on scattered sites. When a house with a white roof is sandwiched in between houses with black roofs, it will not cool as deeply as it would if the roofs on either side were reflective as well. Block preservation is enhanced because of lateral heat transfer. Working block-by-block is clearly the preferred approach with white roof coating. The roof preservation benefits of the white coatings help retain the integrity of all the houses on the block. In addition to reducing indoor temperature, white roof coatings on an entire block or larger area, actually reduce outdoor temperature as well. The evaluation of ECA's pilot program found that outdoor temperature was one degree Fahrenheit lower on the cool block during heat waves. When covering an extended area, white rock coatings can reduce the urban heat-island impact. Cities suffer through the heat of the summer, reaching temperatures five-to-ten degrees higher than the surrounding suburban areas.

In urban neighborhoods, the experience of having preservation and energy conservation work done on every house creates a powerful, shared experience that energizes people on the entire block. We have seen residents on cool blocks come together to tackle other improvement projects such as turning an empty lot into a community garden or planting trees to enhance the cooling effect. White roof coatings lower air-conditioning use and therefore

FIGURES 8.5 and 8.6 (Top) Cool block, Bridesburg neighborhood, Philadelphia. Photograph by ECA staff. Courtesy ECA. (Bottom) Cool block, detail of elastometric roofs. Photography by ECA staff. Courtesy of ECA.

lower peak load, in the process lowering the price of electricity to all customers year-round. Cool roofs slow global warming in two ways: first by reducing electricity use, most of which is coal-fired in the United States. Average air-conditioning load reduction in the Mid-Atlantic region is 20 to 22 percent. For a row house, which has a compact mass and roof area, cooling load reduction is approximately 19 percent.

Furthermore, cool roofs can slow global warming by contributing the reflectance of the earth itself. Increasing the earth's albedo, or reflectance, by converting dark roofs to white, is even more important since the melting of polar ice and of glaciers worldwide is causing highly reflective surfaces of ice and snow to be replaced with dark surfaces of open water and rock and is thereby increasing the rate of melting.[13]

Recent research by Hashem Akbari and Surabi Menon has quantified the effects of white roofs in mitigating climate change.[14] They estimate that replacing a non-reflective dark roof with reflective white roofing on an average house with 1,000 square feet of roof would result in an equivalent CO_2 offset of ten metric tons a year. Applied on a large scale in a number of the world's major cities, white roofs could offset forty-four gigatons of carbon dioxide, more than an entire year of global CO_2 emissions! Heretofore, the only serious strategy to combat global warming has been to reduce greenhouse gas emissions. While that remains the primary imperative, it is now also apparent that we need to increase the earth's reflectance to give us more time to bring down our carbon emissions. The nice thing about white roof coating is that it does both.

The first step toward a "Zero Energy" house is to super-insulate it. With new construction, this is relatively easy. Not so for existing buildings. It is easy enough to super-insulate the attic, which in a row house is really a roof cavity. Because row house roof cavities are so shallow, usually no more than two feet in the front, and sloping to less than a foot in the rear of the building, it is necessary to send in a person to air seal the roof cavity prior to blowing in the insulation. Currently, ECA's preferred material is cellulose, due to many of its properties: it is recycled newsprint, it is relatively environmentally benign, and it is less expensive than the alternatives. Cellulose is also relatively easy to "dense pack," which is necessary in the rear of the cavity, since even a thin person cannot usually get all the way back to do the air sealing.

The challenge is the exterior walls. We need good pilot programs on exterior wall insulation to determine the most cost-effective ways to super-insulate existing brick or masonry walls. A related problem in old buildings that needs to be solved is the mortar joints. In one-hundred-year-old row houses, the mortar is quite worn and is responsible for a great deal of air leakage. Treating this problem by repointing the brick is cost-prohibitive in low-income housing. It may be more cost-effective to treat this problem in a way other than re-pointing the brick. For example, exterior insulation materials, such as spray foam, can be applied over a slurry of binding material, which will restore the structural integrity of the brick. One of the advantages of exterior spray foam is that it can be sculpted to match the original detail of a historic building.

Now is the time for historic preservationists to link more closely with those promoting energy efficiency in buildings. The Obama administration is dedicating considerable resources to energy efficiency in buildings, including low-income houses. This commitment provides the first real opportunity we have had to develop the approaches that will get us to the Zero Energy row house and dramatically increase the energy efficiency of our historic buildings.

Notes

1 For additional information on this and other ECA programs, see *http://www.ecasavesenergy.org*.

2 See *http://www.phila.gov/green/Greenworks/PDFs/GreenworksPlan002.pdf*, accessed 31 March 2010.

3 Energy Information Administration, Office of Energy Statistics, see *http://www.eia.doe.gov/emeu/states/hf.jsp?incfile=sep_use/eu/use_eu_pa.html&mstate=PENNSYLVANIA*, accessed 15 October 2009.

4 New York Sustainable Energy Research and Development Authority, "Evaluation of Energy Star Homes Program," Albany, 2008.

5 Rachael Simmons Schade, *Philadelphia Rowhouse Manual: A Practical Guide for Homeowners* (Philadelphia: Office of Housing and Community Development and City Planning Commission; Washington: National Trust for Historic Preservation, 2008).

6 See *http://www.energystar.gov/index.cfm?c=home_improvements.htm_improvement_hpwes*, accessed 31 March 2010.

7 City of Philadelphia, Department of Public Health statistics, 1999

8 These findings are based on ECA field observation and research.

9 Union of Concerned Scientists, "Climate Change in Pennsylvania: Impacts and Solutions," Cambridge, Mass., October 2008; see *http://www.northeastclimateimpacts.org/pa*.

10 K. E. Wilkes, et al., "Roof Heating and Cooling Loads in Various Climates for the Range of Solar Reflections and Infrared Emitances Observed for Weather Coatings," American Council for Energy Efficient Economy, Washington, D.C., 2000.

11 Michael Blasnik, "Impact Evaluation of the Energy Coordinating Agency of Philadelphia's Cool Homes Pilot Project," Michael Blasnik & Associates, Boston, 5 November 2004.

12 The program recently received a boost from city government; see Sandy Bauers, "Phila. Contest Aims to Whiten City's Roofs," *Philadelphia Inquirer*, 18 February 2010.

13 Tina Kaarsberg and Hashen Akbari, "Cool Roofs Cool the Planet," *Home Energy* 23 (September-October 2006); see *http://www.himeenergy.org*, accessed October 2006.

14 Hashen Akbari et al., "Global Cooling: Increasing World-wide Urban Albedos to offset CO_2," Springer Science and Business Media, New York, 2008.

On the Road Again: Preservation's Urgent Future

Richard Longstreth

In 1980, when the United States was recovering from its second energy crisis in a decade, the National Trust's Preservation Week poster conveyed a vivid message about conservation. A generic "historic" building was fashioned as a gasoline can. The message was clear: preservation equals saving energy. That message got lost in the ebullient years that followed, but, fortunately, it has come back to face us front-and-center, and now with a new sense of urgency.

We cannot afford to squander our built resources any more than we can our natural resources. Sound building stock, with examples grand and humble, large and small, must be retained if we are to mitigate the effects of global warming. To demolish is to negate the energy taken to produce the building in the first place, to expend additional energy on demolition, to transport what has been squandered as detritus to over-used landfills—some of them hundreds of miles away—and then to generate yet more energy in replacing that which

FIGURE 9.1 Demolition of John Bartram Hotel, 233-47 South Broad Street, Philadelphia, 1892, 1895, Angus S. Wade, architect. Photograph by Richard Longstreth, 1967.

was destroyed (fig. 9.1). To preserve, on the other hand, is to recycle on a large scale. We can no longer treat our habitat as a disposable commodity.[1]

The concept of conserving our built environment and appreciating its inherent physical attributes rather than continually replacing it with new components because they are ostensibly better goes against the grain of widespread and long-standing practices in the United States. In 2001, for example, an array of government officials, as well as housing and public policy scholars, applauded what was billed as sweeping plan for the systematic demolition of 14,000 row houses and other buildings as part of Philadelphia's Neighborhood Transformation Initiative. The plan was intended to create large tracts of open land that would appeal to real estate developers. Ultimately the city would benefit from substantial new investments created on a scale that would improve the moribund local economy. Many of the buildings targeted were in an advanced state of deterioration after years of abandonment.[2] As havens for drug dealers as well as rats and other vermin, the derelict dwellings only accelerated the ongoing rate of neighborhood decay, proponents of the plan charged (fig. 9.2). Without massive federal subsidies and with a private sector uninterested in such places unless bolstered by major incentives, what else could municipal leaders do to revitalize the ailing metropolis? The comprehensiveness of the Philadelphia plan, which initially sought to address housing decay on a citywide

FIGURE 9.2 Abandoned row houses, Eighteenth and Jefferson streets, Philadelphia, mid-to-late nineteenth century. Photograph by Richard Longstreth, 1978.

basis, had little precedent nationwide and was among the attributes of the widespread, generally favorable attention it received.

But the less systematic demolition of derelict houses, fueled by large federal, state, and sometimes municipal funds, had in fact been occurring for some years. Practices in Baltimore during the 1990s were not atypical. Abandoned houses in which myriad crimes were committed were razed one by one until many blocks were pockmarked by a profusion of vacant lots, substantially reducing the value of residences still standing in the process.[3] Now, in Baltimore the policy has been reversed, with some ravaged as well as stable neighborhoods being listed on the National Register of Historic Places to stimulate prospective homeowners to take advantage of the state's historic-tax credit. Philadelphia's grand plan never began to approach expectations, and there, too, rehabilitation of its still vast stock of row houses is being encouraged.[4]

Old habits die hard, however. Federal and state housing policies have for the most part ignored preservation—or simply conserving existing housing stock—as a major component of urban revitalization. The momentum remains in new construction—now green construction. In the meantime, the housing stock of cities and towns with depressed economies—from Albany to Buffalo, from Cleveland to Toledo, from Evansville to Gary, from St. Louis to Omaha, continues to deteriorate.[5] Detroit may be the worst example—a city that has become a basket case (fig. 9.3). A city in which most people are reluctant to

FIGURE 9.3 Abandoned and partially dismantled houses, Detroit, early twentieth century. Photograph by Richard Longstreth, 1998.

buy a house—anywhere. A city where home improvement is rare. A city where houses are routinely torched, although the rate has declined. A city where large areas have been completely abandoned and municipal services cut off.[6] But Detroit's problems are shared to a lesser degree by many other places. The course has become tragically predictable. More absentee landlords. Fewer houses occupied by owners who can afford to maintain them. Owners unable to sell their houses. Vacant houses. Abandoned houses. Further disinvestments. Further decline. In 2005, it was estimated that 577 vintage houses were destroyed daily in the United States during the previous three decades.[7] Besides the over six million dwellings thus lost, thousands of decent schools, churches, commercial facilities, parks, and infrastructural components of all kinds—from sidewalks and streets to water and sewer lines—have been laid to waste or became severely underutilized. Most disturbing of all, perhaps, is that the process now occurs in neighborhoods that were considered stable a generation ago (fig. 9.4).[8] The collapse of the real estate market in 2008–09 and the recession generated in part by that crisis have compounded the problem.

FIGURE 9.4 Single- and two-family houses, Joan Street, Cleveland, early twentieth century. Photograph by Richard Longstreth 1995.

The loss of nineteenth- and early twentieth-century residential properties is not just a loss of heritage, but also one of a commodity that is desperately needed: affordable housing. The viability of preservation techniques in rehabilitating houses for persons of moderate income was demonstrated some four decades ago, perhaps most notably by the work of the Pittsburgh History and Landmarks Foundation, but also through projects in Cincinnati, Kansas City, Providence, and a number of other cities.[9] Like energy conservation, most such initiatives lost momentum in the 1980s, but now are coming back to life. Chicago's Neighborhood Housing Services Redevelopment Corporation, with vigorous support from Mayor Richard M. Daley, has targeted the "bungalow belt," comprising about one-third of the city's housing stock, for rehabilitation that is both sensitive to the salient historic attributes of those dwellings and taps into state-of-the-art green technology. The program has become a model, including for the national NeighborWorks © network,

FIGURE 9.5 Vacated commercial center, Hammond, Indiana, most buildings from 1920s and 1930s. Photograph by Richard Longstreth, 1995.

which holds considerable promise for the future.[10]

The vastness of urban decay that has occurred in the commercial as well as in the residential spheres across much of the United States in the past half century—from Hammond, Indiana to Fayetteville, North Carolina—has no comparison in other developed nations (fig. 9.5). Only pervasive poverty or the devastation of war renders such bleakness. And only in the United States is the destruction of once valued landmarks, of former beacons of progress, of great urban symbols such as the former J. L. Hudson department store in Detroit, celebrated as a ritual that will ostensibly bring better times (fig. 9.6).[11] It is a tragedy. It is also a colossal waste. And it is out of the minds of many Americans who simply avoid going to, or thinking about, places they euphemistically dismiss as "inner city."

FIGURE 9.6 J. L. Hudson Company department store, Farmer Street and Gratiot, Grand River, and Woodward avenues, Detroit, 1922-23, 1923-25, 1927-28, Smith, Hinchman & Grylls, architects; demolished 1998. Photograph by Richard Longstreth, 1998.

It is also clear that, however valuable their practices, preservationists need to do a great

FIGURE 9.7 Row houses, off Henry Avenue, Philadelphia, ca. 1910s-20s. Photograph by Richard Longstreth, 1985.

FIGURE 9.8 Overton Park neighborhood, Memphis, developed mostly in early twentieth century. Photograph by Richard Longstreth, 1982.

deal of lobbying to get their message across. Probably most Americans think that the only sustainable buildings are new buildings. A change in outlook is not just necessary among the "experts," but among policy makers at all levels so that the innovative steps that have been taken in a number of communities become standard, widespread practices.

Even when people want to do the right thing and can afford to do so, they may be at a loss as to how to do so properly. Home remodeling is an industry long plagued with unscrupulous players as well as good and decent ones. An enormous amount of harm could occur to our residential fabric alone if there is a new wave of sustainable retrofitting without adequate guidance on the value and viability of historic fabric. No greater service could be done by local, non-profit organizations than to disseminate concise, easy-to-understand handbooks as to basic ways historic buildings can be appropriately brought up to a higher conservation standard. No less important is that such initiatives be done taking a broad perspective on the scope of what is considered historically significant.[12]

Even many run-of-the-mill buildings from the early twentieth century were constructed to last for generations if they are adequately maintained and their systems are periodically upgraded. Those modest row houses that are pervasive in numerous parts of Philadelphia, for example, should be viewed not just on the basis of energy statistics, but also as a precious, renewable resource that would be impossible to duplicated today at anything approaching a reasonable cost. Moreover, they are very significant from a historical perspective, for, among other things, no other metropolis exceeded the City of Brotherly Love into the early twentieth century in the degree of homeownership and in the extent to which the single-family house—owned or rented—was the norm, even for people of moderate means (fig. 9.7).

Sustainability entails far more than buildings. The configuration of streets, sidewalks, alleys, trees, and numerous other components may possess positive environmental attributes in addition to being historical features that give a neighborhood its identity. Many of the houses in a neighborhood such as Overton Park in Memphis are noteworthy examples of domestic architecture from the early twentieth century, but they are but a part of a much richer landscape (fig. 9.8) What would this place be like if all the natural features were taken away? What would be the impact on the environment? Applying the concept of cultural landscape to preservation can be extremely useful in understanding important relationships between components of our

FIGURE 9.9 Grazing and wheat fields, Pottawatomie County, Kansas. Photograph by Richard Longstreth, 1980.

environment over time and in establishing priorities that are holistic and effective.[13] We should look at the landscape—any landscape—not through the dichotomous lens of separating the "natural" and "cultural" worlds, but as closely tied parts of a greater whole.

Historic landscapes—urban and rural—can have ongoing value in traditional and in new ways. Might irrigation systems such as were developed in many Mormon communities during the nineteenth century have a role once again in

FIGURES 9.10 and 9.11 (Top) Rural road, Johnson County, Missouri. Photograph by Richard Longstreth, 1977. (Bottom) Farm complex enshrouded by windbreak, Watonwan County, Minnesota, ca. 1910s–20s. Photograph by Richard Longstreth, 1996.

conserving natural resources?[14] Are the trees that still line many of our first-generation of highways components that should be replaced, either in kind or with other suitable species, as they near the end of their lives? How important is our farmland, not just as a supplier of food, but also as part of an ecological balance and a kind of landscape that is rich in history and local sense of identity (fig. 9.9)? How important are rural roads, not just as reminders of a past fast fading in the rearview mirror, as it were, but also as a network that minimally intrudes on the land (fig. 9.10)? The windbreaks that envelop farmsteads in the parts of the northern Plains states serve a clear function, but they are also signature elements in this landscape that distinguish it from many others (fig. 9.11). Function and history go hand-in-hand in such cases, and even when the particular circumstances change we must bring our ingenuity to bear on these larger, more complex canvases, just as we have done, with remarkable success, for example, in developing adaptive uses for buildings.[15]

FIGURE 9.12 Twinbrook Forest subdivision, Rockville, Maryland, 1952–53, Joseph L. Geerhaert, developer and builder. Photograph by Richard Longstreth, 2008.

The imperatives of sustainability pose a significant challenge to a major component of our landscape that is all too seldom considered positively in this context: that fashioned in large part since World War II and predicated on motorized transportation for movements of almost every kind. What are we to do with the endless tracts of ranch and split-level houses constructed by the hundreds, sometimes by the thousands, in metropolitan areas coast to coast (fig. 9.12)? Environmental historian Adam Rome has argued how wasteful such development was, how much it ruined the natural environment, but what are we to do now with places in which millions live and covet?[16] Assuming the building stock itself is sound, does it make sense to forfeit such places, which represent a major chapter in American history, when millions of people were able to partake in the American dream for the first time—to forfeit them to new, more sustainable development? Or does all the embodied energy in those buildings and in the manipulated landscape of which they are a part and the

FIGURES 9.13 and 9.14 (Top) Reynolds Metals Company executive office building, 6601 West Broad Street, Richmond, Virginia, 1953–58, Skidmore, Owings & Merrill, architects, Charles F. Gillette, landscape architect. Photograph by Richard Longstreth, 2000. (Bottom) River Park, Fourth Street and Delaware Avenue, between N and O streets, S.W., Washington, D.C., 1960–63, Charles M. Goodman, architect, Eric Paepke, landscape architect. Photograph by Richard Longstreth 2003.

energy that would be expended in their demolition, removal, and new site work as well as building construction render such an alternative less appealing? If such places are eradicated, they will be replaced with development that is much denser. Is not the loss of open space—of lawns, shrubbery, trees, and all the rest of elements that have made such places special—a major consequence from an environmental perspective?

Does the pursuit of clean energy sources for private, as well as public, transportation make greater sense than trying to recast a breathtakingly large swath of the settled landscape? The answers are complex, but the point is that while it is easy to cast environmentalist stones at our Levittowns, it is much harder and much more important to seek solutions that have the greatest long-term impact. The well-built postwar tract house may or may not be as long-lasting as the early twentieth-century row house. I do not believe anyone has done the kind of detailed, comparative analysis required for an authoritative comparison. But even if it is not, these places, like the row house and bungalow neighborhoods of a generation or two earlier, are resources we cannot replicate in kind; they are precious, irreplaceable habitat.[17]

The same concerns should be applied to many other components of the post-World War II landscape, created when land was in abundance and a primacy was accorded to open space. A number of office buildings, such as the headquarters of Reynolds Metals in Richmond, Virginia (1953–58), were set on what was then the urban periphery in a park that at once enhanced company prestige and brought pleasure to employees (fig. 9.13).[18] These corporate estates, as they have been called, were like latter-day Medician villas or country houses of the Whig aristocracy—vast, elegant, conceptually adventurous, endowed with an abundance of fine art, and completely engaged with the great parks in which they were set. In some cases, too, as with the Connecticut General Life Insurance Company's offices outside Hartford, Connecticut, their grounds were open to the public.[19]

We can find counterparts in a number of our cities, the result of urban renewal projects, which should not be viewed monolithically and consigned to the historical dustbin, but rather as a wide array of initiatives, some of which have made a significant, lasting contribution to their respective communities.[20] In Washington's Southwest redevelopment area, for example, a galaxy of talented architects and landscape architects created an extraordinarily rich landscape of residential enclaves combining clustered houses with high-rise apartment buildings between the late 1950s and late 1960s.[21] Each of the eight major groups is at once dense and expansive, where copious amounts of open space afford an essential counterpart to the compact arrangement of buildings (fig. 9.14). Additional open spaces connect the housing groups to the waterfront and provide a park for several houses of worship and a public library branch adjacent to the commercial center. No more extensive or fully developed example of modernist urban planning can be found in the United States.[22] Yet two of the city's recent planning directors have dismissed the area as a testament to failure because of the dislocation of low-income residents it perpetrated a half century ago, because it does not conform to more recent models of mixed-use development, and because of all the ostensibly wasted open space. All that space is indeed redundant if one subscribes to the now common practice of planning based primarily on the land's revenue-generating capacity. The approach has obvious advantages for the developers involved and for the municipal tax base from a short-term perspective. But what about creating places in which people want to live? And what about sustainability of a landscape that is primarily built upon and paved? How long will such places remain attractive–for investors and for occupants? Most components of the Southwest redevelopment area were built for the long-term, and their appeal has, if anything, increased with time.

If preservation is good for the cause of sustainability, can it offer us some lessons for new construction? How about the fact that the wood in a standard-issue bungalow of 1910 or 1920 will outlast that in houses erected today for a more affluent clientele? How long will new houses—no matter how high their LEED rating—last? How soon will they have to be replaced because, like a World War II temporary building, maintenance costs exceed value? We can make habitat that is energy efficient, that looks appealing, that is well planned as a domicile and, even, as part of a community. But are we building for generations, or just for the moment? How long will our new, planned communities last? Should we be conserving certain materials to foster longevity in future construction? Can we afford the cost? How should we calculate that cost? It is easy and often very convenient to ignore the long-term costs when most homebuyers accord primacy to price up front. Are dwellings—or buildings of any kind—as disposable products really good investments for their purchasers over time? Is it a good practice for the environment? Should we think again about building for the ages, or at least for a century or two?

However important, such questions are perhaps not those on which preservationists should focus. There is enough on our plate already. Sustainability gives a new imperative to our cause and reinforces the importance of what we do. Preservation is not just for the enjoyment of esthetes and elites. Preservation is not just for tax incentives. Preservation is not just for cultural tourism, nor is it just for all those others reasons we have heard and used to justify what we do again and again. It is for the future of our world. We cannot afford to waste as we have done so prolifically in recent decades. Probably no other living species fouls its own habitat on a regular basis. We cannot replicate what we have lost, but we can reverse the tide and take the steps necessary to conserve a much larger and richer spectrum of our environment than we have in the past. Sustainability affords a new, broad horizon that can greatly reinvigorate preservation if we meet its many challenges. We may have a long way to go before we get there, but we are, indeed, on the road, again. ❁

Notes

I am grateful to Jeffrey Cohen, Howard Gillette, Alisha Lueha, and Laura Spina for supplying me with information that proved invaluable in the preparation of this essay.

1 Among the pioneering studies on the subject, published before global warming became an over-arching issue, are Booz, Allen & Hamilton, *Assessing the Energy Conservation Benefits of Historic Preservation: Methods and Examples*, (Washington, D.C.: Advisory Council on Historic Preservation, 1979); Diane Maddex, ed., *New Energy from Old Buildings* (Washington, D.C.: Preservation Press, 1981) and John A. Burns, *Energy Conservation Features Inherent in Older Homes* (Washington: U.S. Department of Housing and Urban Development and U.S. Department of the Interior, 1982). For a sampling of recent accounts, see Mike Jackson, "Embodied Energy and Historic Preservation: A Needed Reassessment," *APT Bulletin* 36:4 (2005): 47–52; Carl Elefante, "The Greenest Building Is ... One that Is Already Built," *Forum Journal* 21 (summer 2007): 26–38; May Cassar, "Sustainable Heritage: Challenges and Strategies for the Twenty-First Century, *APT Bulletin* 41:1 (2009): 3–11; Richard Moe, "Historic Preservation and Green Building: Finding Common Ground," *Forum Journal* 23 (spring 2009): 9–16; Steve Tilly, "Positioning Preservation in the Center of the Green Arena," *Forum Journal* 23 (spring 2009): 17–23; Baird M. Smith and Carl Elefante, "Sustainable Design in Historic Buildings: Foundations and Future," *APT Bulletin* 40:3-4 (2009): 19–26; and Blair Kamin, "Historic Preservation and Green Architecture: Friends or Foes?" *Preservation* 62 (March-April 2010): 28–33.

2 For background, see Steven J. McGovern, "Philadelphia's Neighborhood Transformation Initiative: A Case Study of Mayoral Leadership, Bold Planning, and Conflict," *Housing Policy Debate*, 17:3 (2006): 529–70. See also Monica Yant Kinney and Clea Benson, "Why City's Housing Started to Fall Down," *Philadelphia Inquirer*, 29 August 2000, A1; Tom Fenick, Jr., "Tearing Down Houses, not Hope," *Philadelphia Inquirer*, 10 March 2002, B1 Larry Eichel, "A Time to Build, A Time to Tear Down," *Philadelphia Inquirer*, 28 April 2002, C5; Rob Gurwitt, "Betting on the Bulldozer," *Governance* 15 (July 2002): 28–30, 32, 34; and Nathan Gorenstein, "City Moves to Seize 803 Lots for Development," *Philadelphia Inquirer*, 2 October 2002, B2.

3 James R. Cohen, "Abandoned Housing: Exploring Lessons from Baltimore," *Housing*

Policy Debate 12:3 (2001): 415–48. See also Tracie Rozhon, "Old Baltimore Row Houses Fall Before the Wrecking Ball," *New York Times*, 13 June 1999, 1, 45.

4 Philip Langdon, "Rowhouse Blues," *Urban Land* 62 (January 2003): 61–64; Kerri Rubman, "Promoting Rehab Tax Credits in Baltimore," *Forum News* 10 (May-June 2004): 3; "Improving Neighborhoods— NTI's Failings," *Philadelphia Inquirer*, 2 June 2008, A14; "Blight Hope," *Philadelphia Inquirer*, 16 October 2009, A1; Rachael Simmons Schade, *Philadelphia Rowhouse Manual: A Practical Guide for Homeowners* (Philadelphia: Office of Housing and Community Development and City Planning Commission; Washington: National Trust for Historic Preservation, 2008).

5 New York City's investment in existing housing stock during this period stands as an important exception.

6 For a sampling of accounts, see Mark Van Allberg, "Property Abandonment in Detroit," *Wayne Law Review* 20 (March 1974): 845–88; Matt Labash, "The City Where the Sirens Never Stop," *Weekly Standard*, 29 December 2008: 16–32; Nancy Kaffer, "Detroit Applies for Federal Funds for Demolition, Rehabbing Work," *Crain's Detroit Business*, 14 December 2009, 1; and Alex P. Kellogg, "Black Flight Hits Detroit," *Wall Street Journal*, 5 June 2010, A1, A10: Some of the state's other depressed cities are following Detroit's lead; see "Demolish Entire Neighborhoods," *Flint Journal*, 27 December 2009, A1.

7 Donovan Rypkema, "Economics, Sustainability, and Historic Preservation," *Forum Journal* 20 (winter 2006): 36.

8 For further discussion, see George Sternlieb and Robert W. Burchell, *Residential Abandonment: The Tenement Landlord Revisited* (New Brunswick, N.J.: Center for Urban Policy Research, Rutgers University, 1973) Barry Bluestone and Bennett Harrison, *The Deindustrialization of America: Plant Closings, Community Abandonment, and the Dismantling of Basic Industry* (New York: Basic Books, 1982); David Wilson and Harry Margulis, "Spatial Aspects of Housing Abandonment in the 1990s: The Cleveland Experience," *Housing Studies* 9:4 (1994): 493–510; and Robert Wilson, "Slash-and-Burn Urban Renewal," *New York Times*, 9 July 1999, A15. The problem is probably the most acute in, but is by no means singular to, cities in the United States; see, for example, Paul Keenan, et al., "Housing Abandonment in Inner Cities—The Politics of Low Demand for Housing," *Housing Studies* 14 (1999): 703–16.

9 See, for example, Arthur P. Ziegler, Jr., *Historic Preservation in Inner City Areas: A Manual of Practice* (Pittsburgh: Allegheny Press, 1971); and Phyllis Myers and Gordon Binder, *Neighborhood Conservation: Lessons from Three Cities* (Washington: Conservation Foundation, 1977). Donovan D. Rypkema, "Historic Preservation and Affordable Housing: The Missed Connection," *Forum Journal* 17 (spring 2003): 5–19, provides a detailed argument for a return to such practices.

10 "The Bungalow Belt," *Urban Land* 62 (January 2003): 65-66; Jack Jensen, "'Green'-habbing Chicago's Signature Bungalows," *Forum Journal* 19 (spring 2005): 27–32; "The National NeighborWorks © Network: Building the Alliance Between Affordable Housing and Historic Preservation," *Forum News* 12 (September-October 2005): 1–2, 6; Eric Wills, "Sweet Home Again," *Preservation* 60 (January-February 2008): 32.

11 Concerning the building, see Richard Longstreth, *The American Department Store Transformed, 1920–1960* (New Haven: Yale University Press, 2010), 17–18; and Michael Hauser and Michelle Weldon, *Hudson's in the Early 1940s* (Charleston, S.C.: Arcadia, 2004). The economic, as well as the ecological, folly of such destruction in the commercial sphere is persuasively argued in Evans Paull, "Quantifying the

Environmental Benefits of the Maryland Historic Tax Credit Program" *Forum Journal* 23 (fall 2009): 12–23.

12 The National Trust provides useful basic information on sustainability; see *http://www.preservationnation.org/issues/sustainability*, accessed 17 February 2010. More detailed guidance for homeowners is afforded by the National Park Service's Technical Preservation Services; see Baird M. Smith, "Conserving Energy in Historic Buildings," *Preservation Briefs* 3 (1978), *http://www.nps.gov/history/hps/tps/briefs/briefs03.htm*, accessed 27 February 2009. Still the best way to reach people who own historic properties is through initiatives undertaken by statewide and local preservation organizations using presentations, newsletters, and websites. While some groups have been proactive in this sphere, much remains to be done.

13 For discussion, see Paul Groth and Todd W. Bressi, *Understanding Ordinary Landscapes* (New Haven: Yale University Press, 1997); Arnold R. Alanen and Robert Z. Melnick, eds., *Preserving Cultural Landscapes in America* (Baltimore: Johns Hopkins University Press, 2000); and Richard Longstreth, ed., *Cultural Landscapes: Balancing Nature and Culture in Preservation Practice* (Minneapolis: University of Minnesota Press, 2008).

14 For historical background, see Richard V. Francaviglia, *The Mormon Landscape: Existence, Creation and the Perception of a Unique Landscape in the American West* (New York: AMS Press, 1978); and L. J. Arrington, "A Different Mode of Life: Irrigation and Society in Nineteenth Century Utah," *Agricultural History* 49 (1975): 3–20.

15 For a thoughtful discussion of the broad subject, see Robert Z. Melnick, "Climate Change and Landscape Preservation: A Twenty-First-Century Conundrum," *APT Bulletin* 40: 3–4 (2008): 35–42.

16 Adam Rome, *The Bulldozer in the Countryside: Suburban Sprawl and the Rise of American Environmentalism* (Cambridge: Cambridge University Press, 2001).

17 For discussion of preservation issues related to the subject, see Richard Longstreth, "The Extraordinary Postwar Suburb," *Forum Journal* 15 (fall 2000): 16–25. For a case study, focusing on planning and landscape issues, see Longstreth, "The Levitts, Mass-Produced Houses, and Community Planning in the United States during the Mid-Twentieth Century," in Dianne Harris, ed., *Second Suburb: Levittown, Pennsylvania* (Pittsburgh: University of Pittsburgh Press, 2010), 123–74, 368–80.

18 For background, see Louise Mozingo, "The Corporate Estate in the USA, 1954-64: 'Thoroughly Modern in Concept, But ... Down to Earth and Rugged'," *Studies in the History of Gardens & Designed Landscapes* 11 (January-March 2000): 25–56; Nicholas Adams, *Skidmore, Owings & Merrill: SOM since 1936* (Milan: Electa, 2007), 88–101; and Jayne Merkel, *Eero Saarinen* (New York: Phaidon, 2006), 68–101.

19 Connecticut General's complex was the subject of a protracted preservation battle; see Stanley Stowe, "A Fight Over, of All Things, an Insurance Office," *New York Times*, 16 July 2000, 14CN; "Shaft for Bunshaft's Modern Box and Noguchi," *Architectural Record* 188 (September 2000): 29; "Fans of Modernism Criticize Cigna's Plans to Raze Offices," *New York Times*, 22 February 2001, B1; "To Save Wilde Still Time to Preserve Cigna's Landmark," *Hartford Courant*, 3 June 2001, *http://www.courant.com/news/opinion/hc-wilde6,0,6249066*, accessed 1 March 2010; Allen Freeman and Kristin Palm, "Old Before Their Time: Monuments of Modernism Struggle to Endure," *Preservation* 53 (July-August 2001): 12, 14; Jane Gordon, "Clash of Cultures Over a Building," *New York Times*, 20 March 2005, 14CN; and Tyler Smith, "How the Wilde Was Won," *Hartford Courant*, 21 May 2006, *http://www.hartfordinfo.org/issues/documents/history/htfd_courant_052106.asp*, accessed 1 March 2010. The building was preserved, but the

landscape, which was an integral and vibrant component of the scheme, was markedly altered for a golf course.

20 Richard Longstreth, "The Difficult Legacy of Urban Renewal," *CRM: The Journal of Heritage Stewardship* 3 (winter 2006): 6–23.

21 Richard Longstreth, "Brave New World: Southwest Washington and the Promise of Urban Renewal," in Longstreth, ed., *Housing Washington: Two Centuries of Residential Development and Planning in the National Capital Area* (Chicago: Center for American Places, and University of Chicago Press, 2010), 255–80, 353–65.

22 For background, see Eric Mumford, *Defining Urban Design: CIAM Architects and the Formation of a Discipline, 1937–69* (New Haven: Yale University Press, 2010).

CONTRIBUTORS

Shary Page Berg, FASLA, has been a leader in the evolving field of landscape preservation for over thirty years. She worked for the National Park Service for ten years before establishing her own landscape preservation practice in 1987. She works primarily with public and non-profit clients, including the Massachusetts Heritage Landscape Inventory Program, is an adjunct assistant professor in the Master of Arts; Historic Preservation Master's Program at Goucher College, and a past president of the Alliance for Historic Landscape Preservation.

Rebecca Howell Crew is an architectural historian at Parsons-Brinckerhoff in Baltimore. Educated at Randolph-Macon Women's College and the University of Virginia, she has experience documenting a variety of cultural landscapes and in conducting historic preservation compliance projects, particularly for transportations agencies.

Karana Hattersley-Drayton is the historic preservation project manager in Fresno's Planning and Development Department. She has served on the board of directors of the Vernacular Architecture Forum and on the California State Historic Resources Commission. She has a B.A. and M.A. and has completed three years of coursework toward the Ph.D. in architectural history at the University of California, Berkeley.

Meisha Hunter has been researching historic waterworks since 1996. In 2006 she was nominated for the Rome Prize in Historic Preservation from the American Academy in Rome. Subsequently, she pursued post-graduate studies in preservation at Columbia University. Her current work entails assessing global climate change's potential impact on waterworks heritage, establishing a working group to write preservation management guidelines, and serving on a preservation planning team for an ancient aqueduct in Spain. She is a senior preservationist at Li/Saltzman Architects in New York.

Thomas F. King is best known for his work with indigenous groups and local communities, using historic preservation laws to insist that their cultural concerns be considered in planning. He is the author of eight books on heritage/cultural resource management through Left Coast Press and Altamira Press as well as many journal articles, popular articles, and internet offerings on heritage topics. His most recent nonfiction book (*Our Unprotected Heritage*) is a critique of contemporary cultural resource management and environmental impact assessment, with recommendations for improvement. He publishes a blog at *http://crmplus.blogspot.com/*.

Valencia Libby is a landscape historian living in New England. She holds degrees in landscape architecture from Cornell and in public horticulture from the University of Delaware. For the past twenty years she has taught landscape architecture and horticulture at Temple University and in 2004 she received a Fulbright Distinguished Chair in Landscape History in Portugal. Professor Libby has been a consultant for the historic landscapes of Colonial Williamsburg, the gardens at Winterthur and Gibraltar in Delaware, and Oxmoor Plantation in Kentucky.

Richard Longstreth is professor of American studies and director of the Graduate Program in Historic Preservation at George Washington University. A past president of the Society of Architectural Historians, he has written extensively on historical and preservation subjects. His most recent books include *The American Department Store Transformed, 1920–1960* (2010) and two edited volumes: *Housing Washington: Two Centuries of Residential Development and Planning in the National Capital Area* (2010) and *Cultural Landscapes: Balancing Nature and Heritage in Preservation Practice* (2008).

Liz Robinson has served as the executive director of the Energy Coordinating Agency of Philadelphia (ECA) since its founding in 1984. ECA provided more than 86,000 energy services to over 40,000 low-income households last year, including energy conservation treatments for more than 5,000 households. Ms. Robinson is a leader in the development of energy-efficient housing, sustainable building practice, and solar energy in Pennsylvania. She also has played an instrumental role in disseminating cool roof coating nationally. She has written articles for numerous trade publications, including *Energy Design Update*, *Home Energy Magazine*, and *GRID*.

Richard Wagner is director of the Master of Arts in Historic Preservation Program at Goucher College and principal in the Baltimore architectural firm, David H. Gleason Associates, where he specializes in historic preservation and rehabilitation projects. He has written extensively on preservation and commercial district revitalization. Professor Wagner is a member of the Maryland Advisory Council on Historic Preservation and member of the board, College Park City/University Partnership, a local development corporation. He received his B. Arch. from the University of Virginia and Ph.D. from the University of Edinburgh.